AS Level Maths is Really Hard

AS level maths is seriously tricky — no question about that.

We've done everything we can to make things easier for you. We've scrutinised past paper questions and we've gone through the syllabuses with a fine-toothed comb. So we've found out exactly what you need to know, then explained it simply and clearly.

We've stuck in as many helpful hints as you could possibly want — then we even tried to put some funny bits in to keep you awake.

We've done our bit — the rest is up to you.

What CGP's All About

The central aim of Coordination Group Publications is to produce top quality books that are carefully written, immaculately presented and astonishingly witty — whilst always making sure they exactly cover the syllabus for each subject.

And then we supply them to as many people as we possibly can, as cheaply as we possibly can.

Contents

Section One — Functions

Section Two — Sequences and Series

Section Three — Powers, Logs and Exponentials

Section Four — Differentiation

Section Five — Integration

Section Six — Numerical Methods

Section Seven — Trigonometry

This book covers all the main syllabuses: Edexcel, AQA A & B, OCR A & B.
There's a note at the top left of each double page to tell you if there's a bit you can ignore.

Published by Coordination Group Publications Ltd
Typesetters:
Martin Chester, Sharon Watson
Contributors:
Bill Dolling, Dave Harding, Lin McIntosh, Claire Sing,
Mike Smith, Caroline Starkey, Kieran Wardel
Editors:
Charley Darbishire, Simon Little, Tim Major and Andy Park
Many thanks to Glenn Rogers *for proofreading.*

ISBN 1 84146 987 4
Groovy website: www.cgpbooks.co.uk
Jolly bits of clipart from CorelDRAW
Printed by Elanders Hindson, Newcastle upon Tyne.

Functions

Skip these two pages if you're doing AQA B.

Thank you for choosing the CGP P2 book. Used correctly, this product should cater for all your P2 revision needs.

A Function changes One Number into Another Number

1) Basically, functions take one number and 'map' it onto another number. (For example, the function 'add 1' would map 3 onto 4.)
2) A function has a domain and a range. The domain is just the set of numbers that you can 'input' into a function. The range is the set of possible 'outputs'.
3) Each number in the domain is mapped onto only one number in the range.
4) Graphs show mappings. Each number on the x-axis (domain) is mapped onto another number on the y-axis (range).

The set of numbers you begin with is called the **domain** — these go on the **x-axis**.

The set of numbers you finish up with is called the **range** — these go up the **y-axis**.

Remember that 'domain' comes before 'range' in the alphabet — like x comes before y.

5) Very often, a function's domain is the set of real numbers $\mathbb{R}$ (or maybe just part of it).
6) The set of real numbers is basically all 'normal' numbers, e.g. 0, 12, -65, 0.341876234, $\frac{1}{3}$, π, etc.
7) If x is a real number, then you can write $x \in \mathbb{R}$. The $\in$ symbol means 'belongs to the set'.

Example: A function is defined by $f(x) = x^3 + 3$, $x \in \mathbb{R}$, $-2 \le x \le 3$. Find the range.

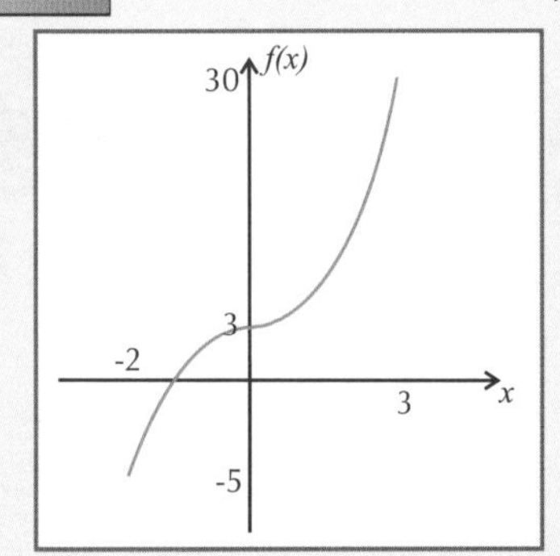

The bit in red above is the domain — it tells you what values x can take.

Here x can be any real number between -2 and 3 (inclusive).

To find the range, it's best to draw a graph. This one's a normal cubic shape.

$f(-2) = -8 + 3 = -5$ and $f(3) = 27 + 3 = 30$, and the graph doesn't do anything too weird between these points (like shoot off to y = 1000, for example).

So the range is $-5 \le f(x) \le 30$, $f(x) \in \mathbb{R}$.

A One-to-One Function is Different from a Many-to-One Function

The example below is a one-to-one function — different values in the domain give different values in the range.

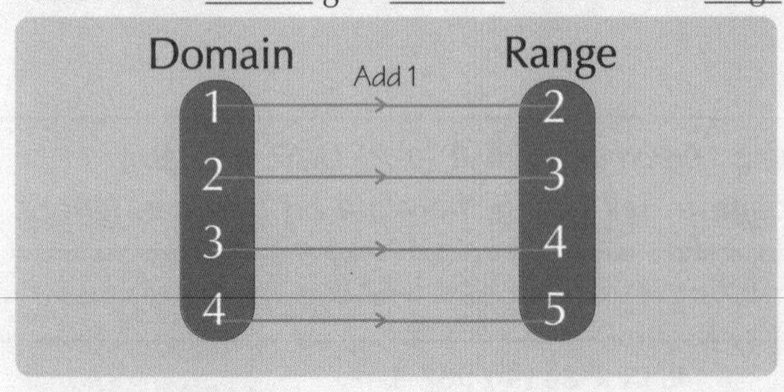

This is a many-to-one function — different values from the domain can give the same value in the range.

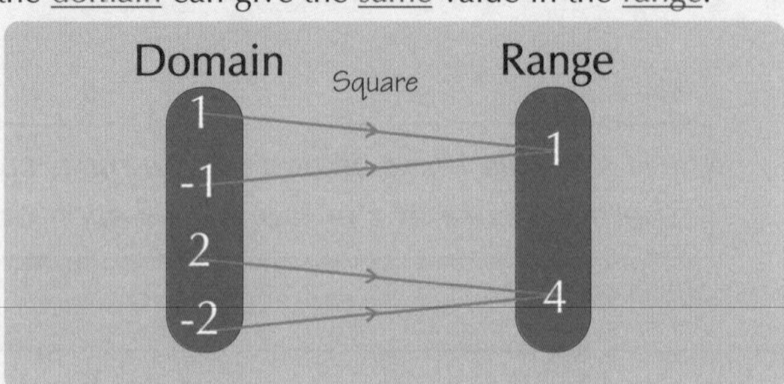

Example:

$f(x) = x^2$

$x \in \mathbb{R}$

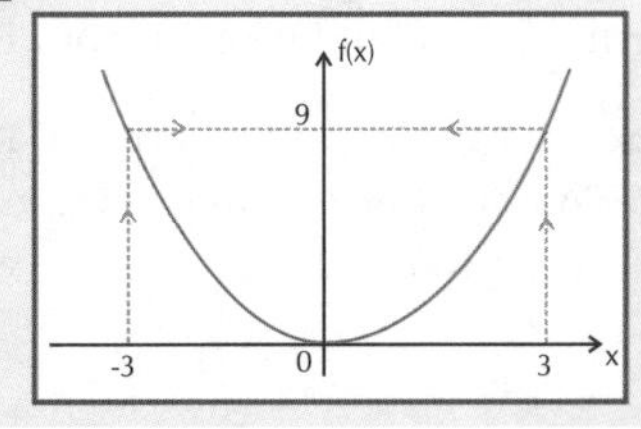

$f(3) = 9$ and $f(-3) = 9$

This is a many-to-one function because 'many' (i.e. more than 1) values in the domain give the same value in the range.

Functions

The definitions of even and odd functions make this topic look more difficult than it actually is — so it's probably easier to remember what the pictures look like. (A picture speaks a thousand words, and so on and so on...)

Even Functions and Odd Functions Have Different Kinds of Symmetry

For an odd function: $f(x) = -f(-x)$

For an even function: $f(x) = f(-x)$

Graphs of odd functions have rotational symmetry of order 2 about the origin.

Example: Is the graph of $y = \sin x$ odd, even or neither?

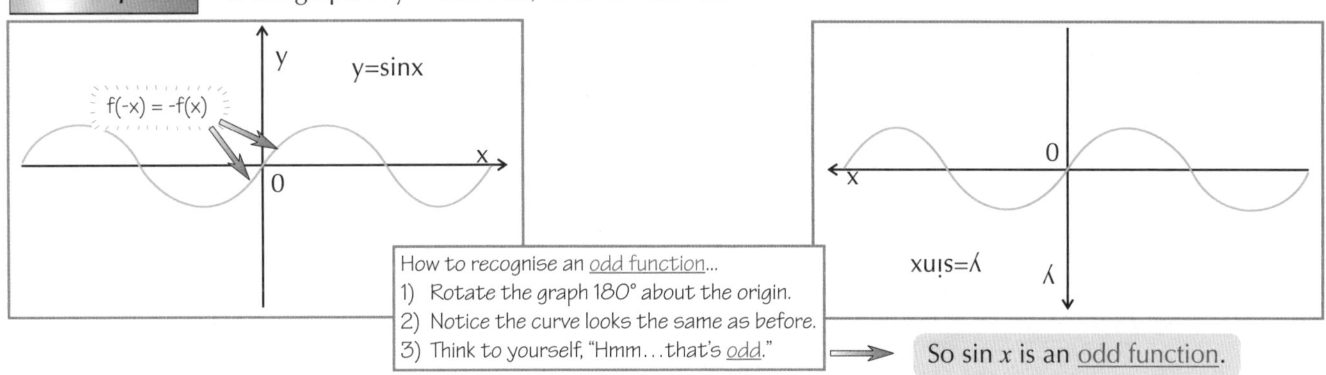

So $\sin x$ is an odd function.

Even functions have reflection symmetry in the y-axis — i.e. even functions have a 'mirror line' on the y-axis.

Example: Is the graph of $y = 4x^2 - 9$ odd, even or neither?

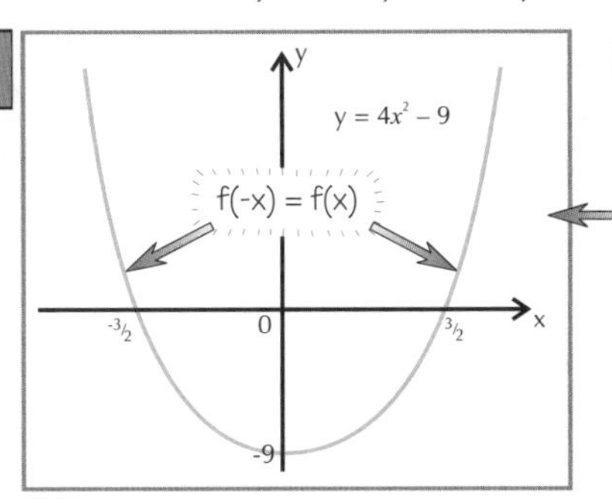

The graph has reflection symmetry in the y-axis, so $y = 4x^2 - 9$ is an even function.

Remember: **O**dd – r**O**tation **E**ven – r**E**flection

But a function doesn't have to be odd or even — it can be neither.

Practice Questions

1) ***Copy the table and tick the correct box for each function.***

Function	$y = x^2 + 3$	$y = 1/x$	$y = x^2 + 2x + 3$	$y = 2x^3$
Even				
Odd				
Neither				

2) ***Sample Exam question:***

The function $f(x)$ is given by $f(x) = x^2 - 2$, where $x \in \mathbb{R}$.

a) (i) What is the range of $f(x)$? [2 marks]

(ii) Is $f(x)$ an even function, an odd function or neither? Explain your answer. [1 mark]

b) The function $g(x)$ is given by $g(x) = f(x) - 3x$.
Is $g(x)$ an even function, an odd function or neither? Explain your answer. [1 mark]

Come on admit it — for an AS Maths book, that wasn't such a bad start...

You put a number into a function, and you get another number out — easy. Remember — each number in the domain gets mapped onto just one number in the range. However, there are functions where more than one number in the domain gets mapped to the same number in the range. But even that's not enough to make this topic too horrendous.

More Functions

Skip these two pages if you're doing AQA B.

Composite (or compound) functions are loved by examiners the world over. But they're not hard. You put a number in one function, and put the result into a second function. That's all there is to it.

For a **Composite Function** Start with the **Letter Nearest the x**

If you have two functions (called f and g, say), then $fg(x)$ means you take x, 'do g to it', and then 'do f to the result'.

Example: $f(x)=3x+1$, $g(x)=x^2$ where $x\in\{1,2,3\}$. Find the ranges for (i) $fg(x)$ and (ii) $gf(x)$.

(i) fg(x) means 'do g, then f'.

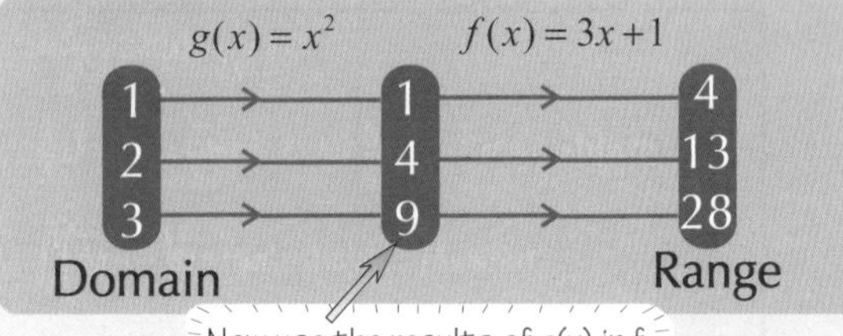

Now use the results of g(x) in f.

(ii) gf(x) means 'do f, then g'.

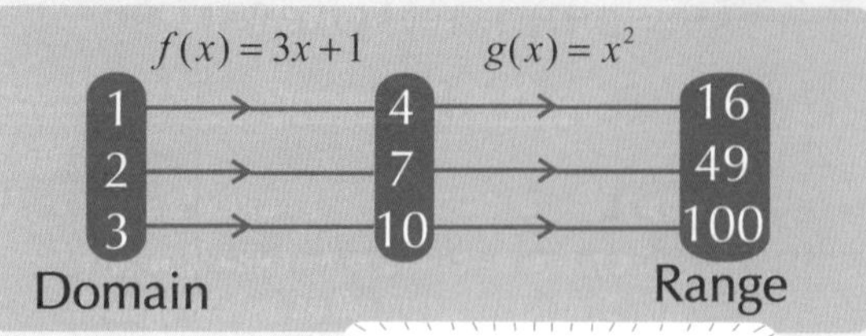

Use the results of f(x) in g.

So, the ranges are (i) $\{4,13,28\}$ and (ii) $\{16,49,100\}$.

The same domain gives different ranges for fg and gf. So fg is definitely not the same as gf.

Remember — normally, $gf(x)\neq fg(x)$.

Example: With f and g as above, write down an expression for (i) $fg(x)$ and (ii) $gf(x)$.

(i) $fg(x) = f[g(x)] = f[x^2]$

But f means 'multiply by 3 and add 1'.

So $f[g(x)]=3(x^2)+1=3x^2+1$

(ii) $gf(x) = g[f(x)] = g[3x+1]$

But g is the 'squaring' function.

So $g[f(x)]=(3x+1)^2=9x^2+6x+1$

An **Inverse Function** Gets You Back to where you **Started**

The inverse function of $f(x)$ is written $f^{-1}(x)$. You can find inverse functions using this 3-step method:

Example: If $f(x)=4x+1$, find $f^{-1}(x)$.

Write $y=4x+1$

1) Replace y with x and x with y. $x=4y+1$

2) Make y the subject. $y=\frac{x-1}{4}$

3) Write $f^{-1}(x)$ instead of y. $f^{-1}(x)=\frac{x-1}{4}$

Don't get confused with the notation:
1) $f^{-1}(x)$ doesn't mean the reciprocal, $\frac{1}{f(x)}$.
2) And $f^{-1}(x)$ definitely doesn't mean the derivative $f'(x)$.

The range of the function is the domain of its inverse, and vice versa.

Example:

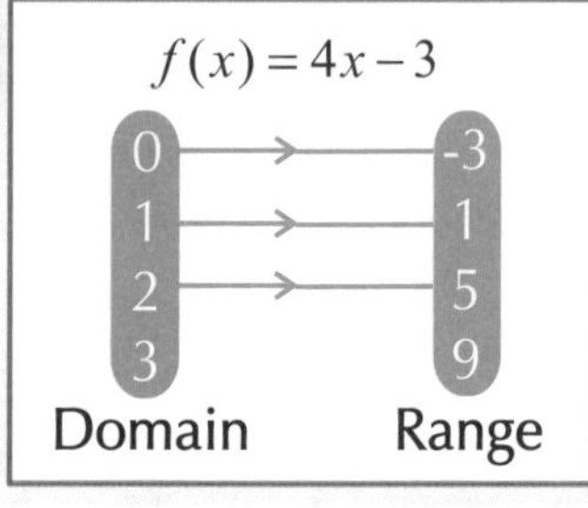

Work out the inverse function:

$y = 4x - 3$

1) Swap x and y: $x = 4y - 3$

2) Rearrange: $y=\frac{x+3}{4}$

3) Swap y for $f^{-1}(x)$: $f^{-1}(x)=\frac{x+3}{4}$

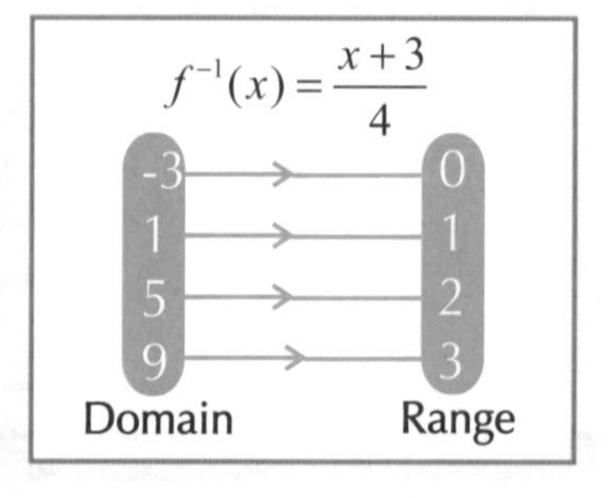

It even works for those 'hard' trig functions.

Example: $f(x)=\cos 3x$ for $0\leq x\leq\frac{\pi}{3}$. What is the range of $f^{-1}(x)$?

Answer: $0\leq f^{-1}(x)\leq\frac{\pi}{3}$.

More Functions

Only One-to-One Functions have Inverses

1) A function can only map a number in the domain onto one number in the range.
2) This means that you can't find an inverse function of a many-to-one function, since the inverse would need to map one number onto two or more different numbers.

Example: $f(x) = x^2$ ($x \in \mathbb{R}$) is a many-to-one function — for example, both 3 and -3 are mapped onto 9.

This means that you can't find an inverse of f — since 9 could go back to either 3 or -3.
(Taking square roots always gives you two possible answers — a positive one and a negative one.)

Remember: If $x^2 = 9$, $x = \pm 3$.

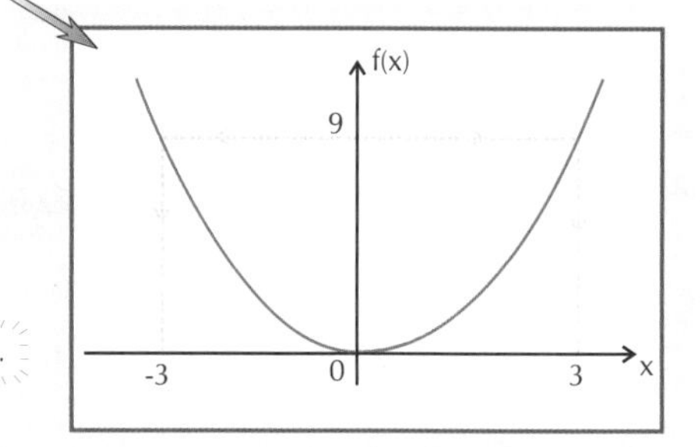

3) However, you can sometimes get around this by 'restricting' the domain.
If you use just enough numbers in the domain, you can get a one-to-one function. Then you can find an inverse.

Example: $f(x) = x^2$, but only for $x \geq 0$. This means you get:

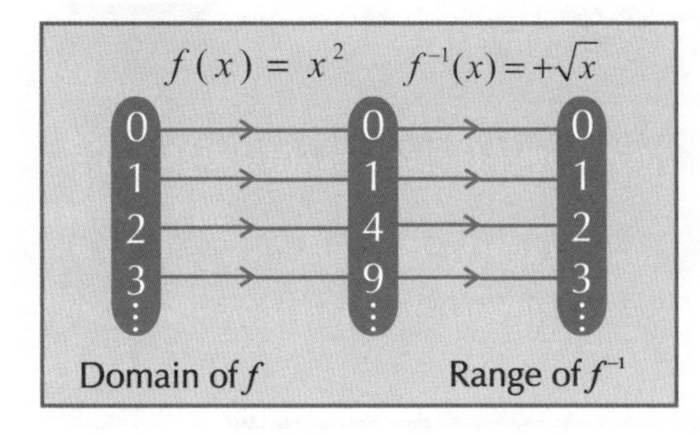

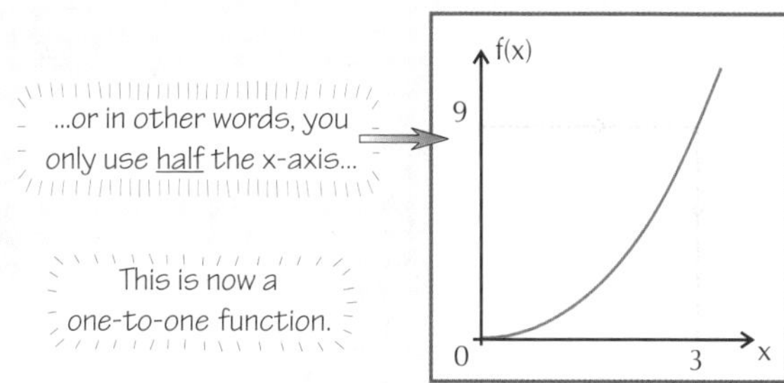

Example: $f(x) = (2x+3)^2$, $x \in \mathbb{R}$, $x \geq -\frac{3}{2}$. Find $f^{-1}(x)$, stating its domain.

Find the inverse in the normal way — write $y = (2x+3)^2$, then swap x and y and rearrange for y:

$f^{-1}(x) = \frac{\sqrt{x} - 3}{2}$. The domain of $f^{-1}(x)$ is the range of f — this is $x \in \mathbb{R}$, $x \geq 0$.

2x + 3 can be any non-negative value, and so f(x) can also be any non-negative value.

Practice Questions

1) *If $f: x \to 2 - x$ and $g: x \to x^2 + 1$, find a) $fg(3)$, b) $ff(5)$, c) $gf(1)$*
2) *The functions g and h are defined as follows: $g(x) = 3x + 4$ and $h(x) = 2x^2$*
   Find the single functions a) $hg(x)$, b) $gh(x)$, c) $hh(x)$.
3) *If $h(x) = x^2 - 5$, $x \in R$, $x \geq 0$, find $h^{-1}(x)$, stating its domain.*
4) ***Sample Exam question:***

The functions $f(x)$ and $g(x)$ are given by: $f(x) = x^3$ ($x \in \mathbb{R}$) and $g(x) = x - 2$ ($x \in \mathbb{R}$). Find

a) (i) $f(-2)$, (ii) $fg(-2)$, (iii) $gf(-2)$ [3 marks]

b) Write an expression for (i) $fg(x)$, (ii) $gf(x)$. [4 marks]

c) If $h(x) = \frac{x^2 - 2}{4}$, (for $x \in \mathbb{R}$, $x \geq 0$), find $h^{-1}(x)$, stating its domain and range. [4 marks]

I know what you're thinking — I'm an AS Maths student, get me out of here...

With composite functions, remember to start with the function nearest the x — which means going from right to left. (Yep, mathematicians read backwards — I knew they were weird.) As for inverses, remember the 3-step method and the trick for turning a many-to-one function into a one-to-one function. Then, as Shakespeare wrote, Bob's your uncle.

Graphs of Functions and Their Inverses

Skip these two pages if you're doing AQA B.

You'll need to sketch graphs of functions and their inverses on most Exam papers. No problem.

Sketching graphs is easy — but use the *Right Domain*

Example: Draw the graph of f: $x \to 3x+2,\ x \in \mathbb{R},\ 1 \le x \le 5$.

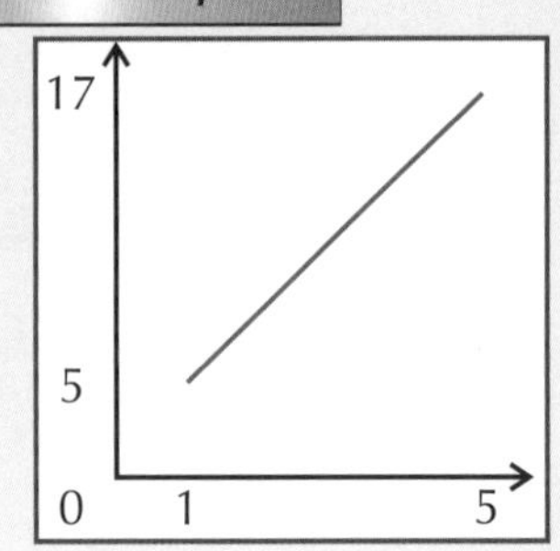

f: x → 3x + 2 means the same as f(x) = 3x + 2.

The domain only goes from 1 to 5, so only draw the line between x = 1 and x = 5.
So you only need y-values from 5 to 17.
Don't be tempted to draw the line for all values of x.

Example: Draw the graph of $f(\theta) = \tan 2\theta$ for $-\frac{\pi}{4} \le \theta \le \frac{\pi}{4}$.

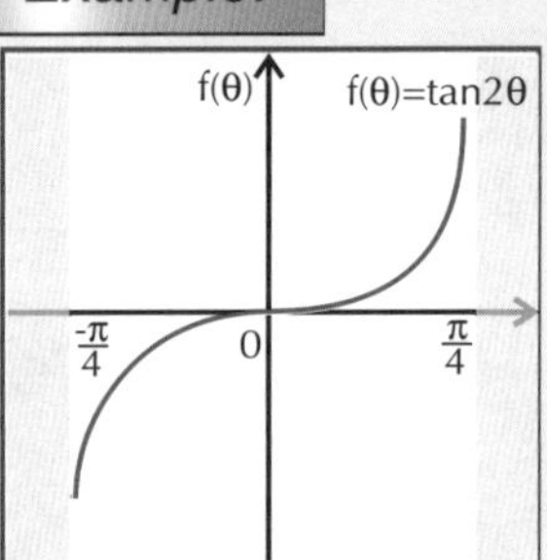

Shade the parts of the graph you're not going to use — then you won't get carried away and draw 'too much graph'.

The Graph of an *Inverse* Function is a *Reflection* in the Line *y = x*

There's nothing complicated about the graph of an inverse function — it's just the reflection of the graph you started with. The 'mirror line' is the line y = x.

Example: If $f(x) = x^3$, sketch $f^{-1}(x)$.

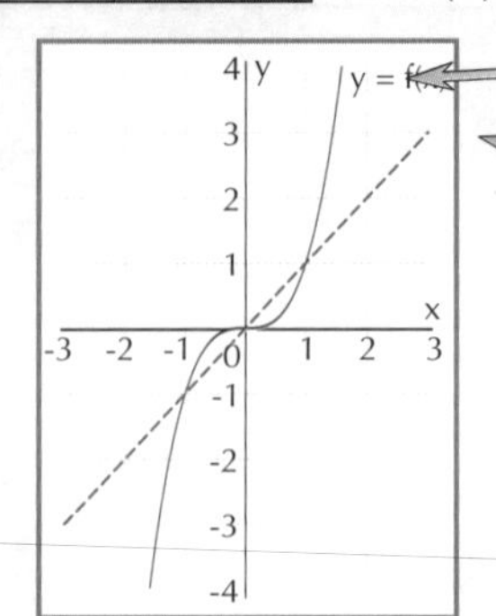

f(x) is a cubic function, so its graph looks like this...

The 'mirror line' is y = x...

...so if you reflect the graph of f(x) in the mirror line...

...you get the graph of $f^{-1}(x)$, like this:

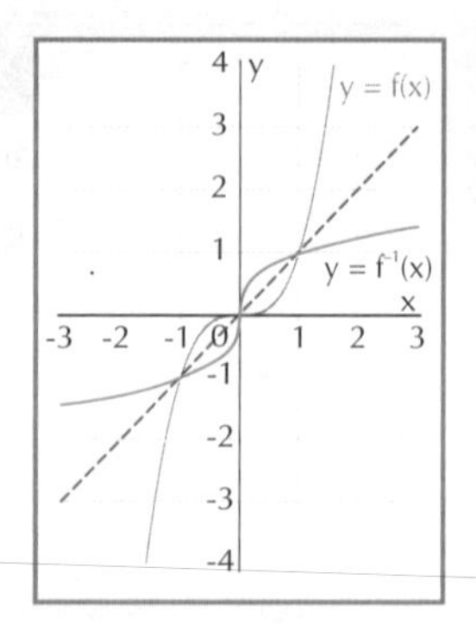

This is the graph of the inverse function.

You can check your inverse function graphs like this...
Copy your original graph onto paper.
Rotate your page 90° anticlockwise.
Then turn your paper over like you're turning the page in a book.
If you hold the page up to the light, then that's what your inverse graph should look like.

Graphs of Functions and Their Inverses

Have a look at this **Worked Exam-Style Question**

For $f(x)=x^2-6x+5$, $x\in\mathbb{R}$, $x\geq 3$
(a) find the range of f,
(b) write down the domain and range of f^{-1}
(c) find $f^{-1}(x)$
(d) sketch the graph of $f^{-1}(x)$, indicating clearly where the curve meets the axes.

Solution:

a) Sketching quadratics is P1 work.
A sketch of f(x) shows that the minimum value is -4, when x = 3.

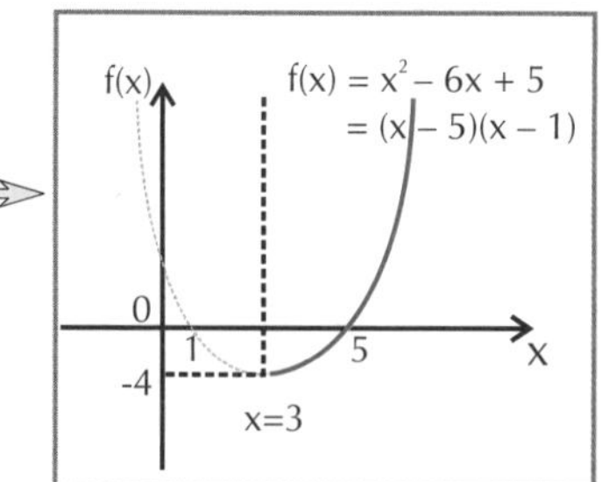

Now look at the y-axis for the range: So, $f(x)\geq -4, f(x)\in\mathbb{R}$

b) Easy game.
Domain of $f^{-1}(x)$ is $x\geq -4$, $x\in\mathbb{R}$
Range of $f^{-1}(x)$ is $f^{-1}(x)\geq 3$, $f^{-1}(x)\in\mathbb{R}$

Remember:
Range of f(x) = domain of $f^{-1}(x)$.
Domain of f(x) = range of $f^{-1}(x)$.

c) Looks a bit trickier, but just follow the rules...
Write $y=x^2-6x+5$, and then swap x and y to get: $x=y^2-6y+5$

Now you have to rearrange to get y in terms of x. It's a quadratic, so complete the square...
$x=(y-3)^2-4$ ⟸ Since $(y-3)^2=y^2-6y+9$.
So $y-3=\sqrt{x+4}$ and therefore $y=3+\sqrt{x+4}$

We're only interested in the positive square root here, so that we stay within the range of f^{-1} (or the domain of f).

The inverse of f is: $f^{-1}(x)=3+\sqrt{x+4}$ $x\in\mathbb{R}, x\geq -4$

d) If you think back to your graph-sketching days, this should be easy. If not, you'll have to think a little more.

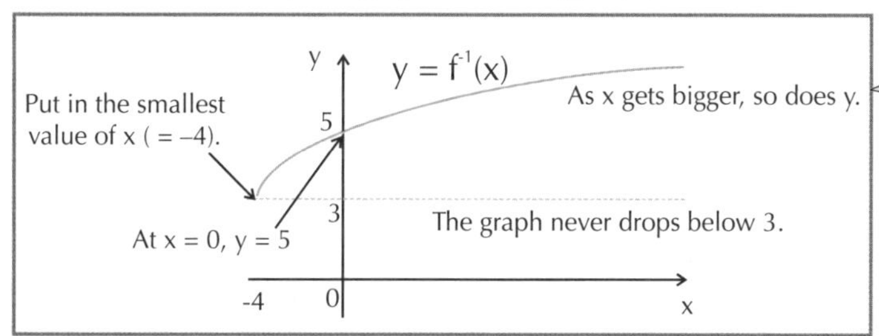

You should check that this is a reflection of your graph from part a) in the line y = x — it is, so we're okay.

A quicker way would be to remember the basic shape of the graph $y=\sqrt{x}$, then move it 4 to the left and 3 up.

Practice Question

1) For $g(x)=(x+2)(x-4)$, $x\in\mathbb{R}, x\geq 1$,
(a) find the range of g. [2 marks]
(b) find $g^{-1}(x)$, stating the domain and range. [3 marks]
(c) sketch the graph of $g^{-1}(x)$, indicating clearly any points where the curve meets the axes. [3 marks]

As my Nan would say, it's as dry as a bunch of old sticks...

Exam questions have a habit of looking dead scary at first glance. The one on this page might have looked worryingly hard a while back, but learn the lingo and the little tricks, and they usually look a whole lot less intimidating. That's the good news. The bad news is that while this topic isn't the hardest in the world, it's a bit dry and dusty. Still needs learning, though.

Curve Sketching

Skip these two pages if you're doing OCR A.

It's dead important to be able to sketch curves — in fact, it's best if you can do them without thinking. This is because: a) you get whole questions just on curve sketching, and b) sketching curves can help in loads of other questions.

Graphs of **High Powers of x** are a piece of cake

You have to be able to sketch graphs of $y = x^n$, where n is a positive or negative whole number. Fortunately, this is easy if you remember a few basic patterns.

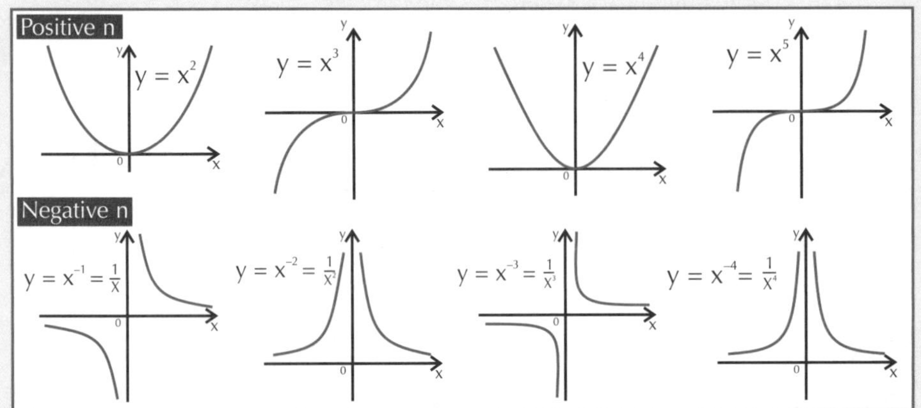

For even values of n, $y = x^n$ is an even function (see page 3).

For odd values of n, $y = x^n$ is an odd function.

As n increases, the graphs get steeper.

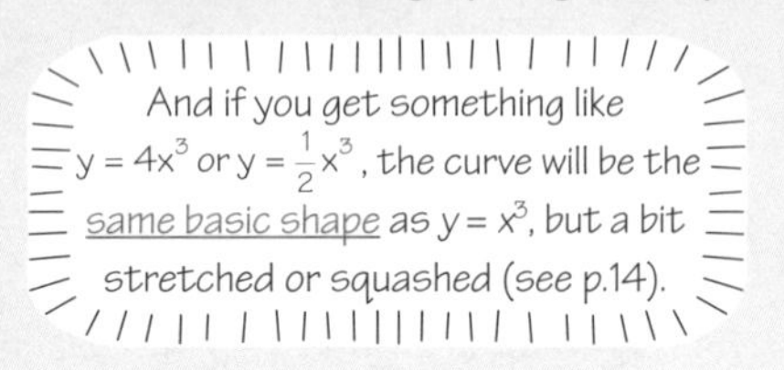

A **Sketch** of a curve contains **Just the Important Bits**

If you need to sketch a graph, just show the important stuff:

1) **general shape**,
2) **intersection with axes**,
3) **turning points**,
4) **what happens as x gets very big**.

Example: Sketch $y = x^2 - 3x - 4$

1) It's a quadratic function, so the graph is a u-shaped parabola.
2) Find where it cuts the x-axis by factorising it or by using the quadratic formula. This one factorises easily: $y = (x-4)(x+1)$. So $y = 0$ when $x = 4$ and $x = -1$.
3) It cuts the y-axis when $x = 0$ — which gives $y = -4$.
4) There's a turning point at $x = 1.5$ (halfway between $x = -1$ and $x = 4$). Here, $y = -6.25$

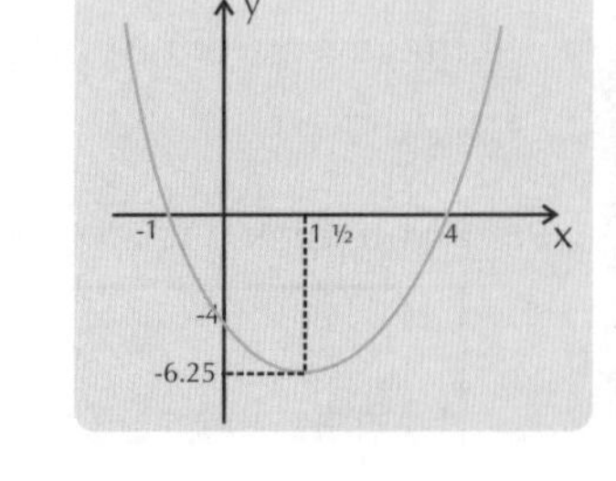

If you're doing **OCR B** you also need to be able to sketch $y = 2^x$.

1) When x = 0, y = 2^0 = 1 — so the curve cuts the y-axis at (0, 1).
2) Now work out what happens when x gets very big positive: $2^1 = 2$, $2^2 = 4$, $2^3 = 8$, $2^4 = 16$, $2^5 = 32$, etc., and so as x gets bigger and bigger, 2^x also gets bigger and bigger (more and more quickly). In maths-speak, you write 'as $x \to \infty$, $y \to \infty$' (and say, "as x tends to infinity, y tends to infinity").
3) Now think about what happens when x gets very big negative: $2^{-1} = 0.5$, $2^{-2} = 0.25$, $2^{-3} = 0.125$ etc., and so as x gets more and more negative, 2^x gets smaller and smaller but stays positive. Write 'as $x \to -\infty$, $y \to 0$' ("as x tends to minus infinity, y tends to zero").

Now sketch the graph...

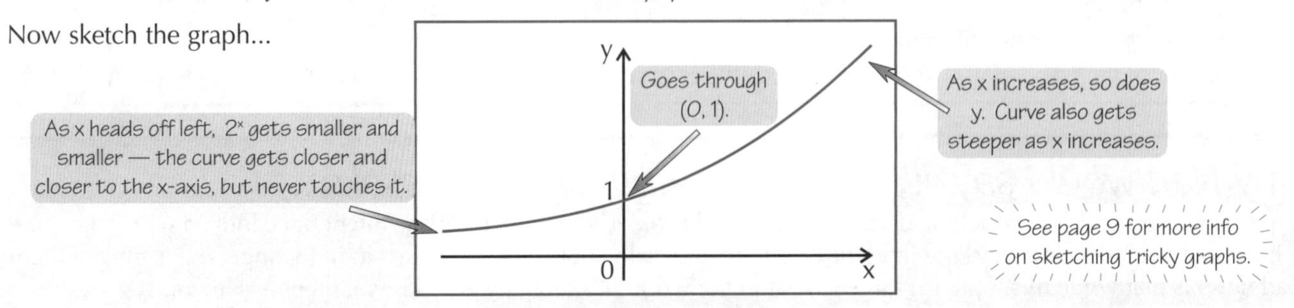

See page 9 for more info on sketching tricky graphs.

Curve Sketching

Use this 3-Step Method for Sketching Graphs — including Asymptotes

With $y=\frac{1}{x}$, the graph gets closer and closer to the x-axis as x gets very big, but it never crosses it. Similarly, it gets really close to the y-axis for really small x, but never meets that either. The x- and y-axes are called asymptotes. Other graphs might have different asymptotes.

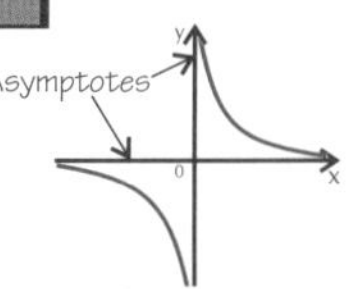

Example: Sketch the graph of $y=\frac{-2}{(2x+3)}$ $\quad x\in\mathbb{R},\ x\neq-\frac{3}{2}$.

You can't divide by zero, so the function is undefined at $x=-\frac{3}{2}$. You always get asymptotes where the function is undefined.

Stage 1 — Deal with the axes and find any asymptotes

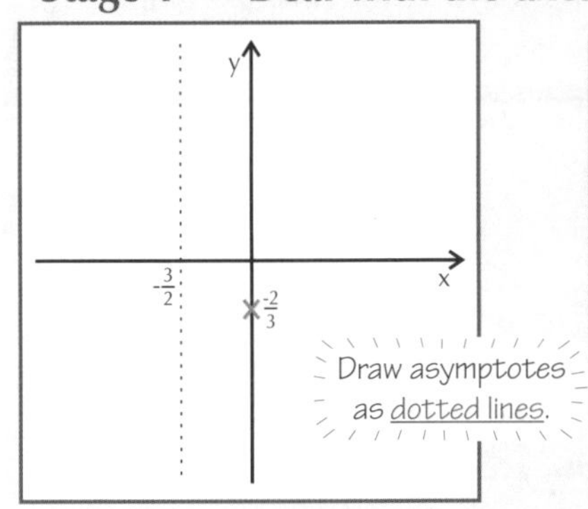

Draw asymptotes as dotted lines.

1) You always get an asymptote where the bottom line of the function is zero (the function's undefined there) — this is at $x=-\frac{3}{2}$.
2) When x = 0, $y=-\frac{2}{3}$ — this is where the curve crosses the y-axis.
3) The curve crosses the x-axis when y = 0. To find where y = 0, you make the top line of the fraction equal to zero (if you can). Here, the top line is -2 (which is never equal to zero), so the curve never meets the x-axis.

Stage 2 — Decide what happens as x gets very big and as it approaches any asymptotes

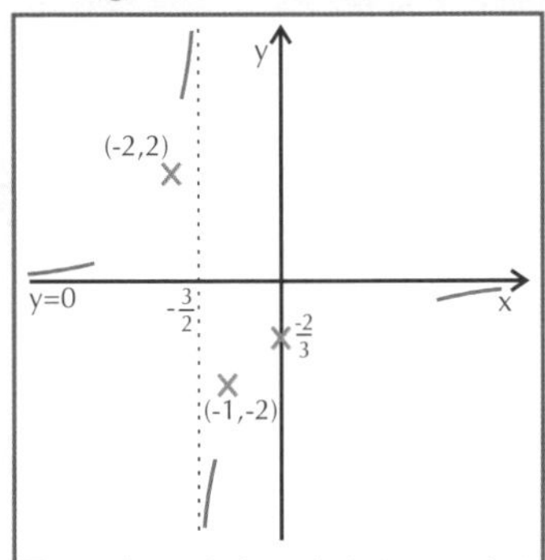

1) When x is very big and positive, (i.e. as $x\to\infty$),
$$y=\frac{-2}{\text{something very big and positive}}=\text{something very small and negative, i.e. } y\to 0.$$
2) Similarly, when x is very big and negative, (i.e. as $x\to-\infty$), $y\to 0$ but is positive.
3) When x is near $-\frac{3}{2}$ (but bigger than $-\frac{3}{2}$), the bottom line is near zero but positive, so
$$y=\frac{-2}{\text{something very small and positive}}=\text{something very big and negative, i.e. } y\to-\infty.$$
4) Similarly, when x is near $-\frac{3}{2}$ (but smaller than $-\frac{3}{2}$), $y\to+\infty$.

Find coordinates of points near the asymptotes to check — e.g. (-1, -2) and (-2, 2).

Stage 3 — Fill in the gaps using a bit of common sense

Look at the information you've already got and join up the bits into some kind of smooth curve (but expect a jump either side of any asymptotes).

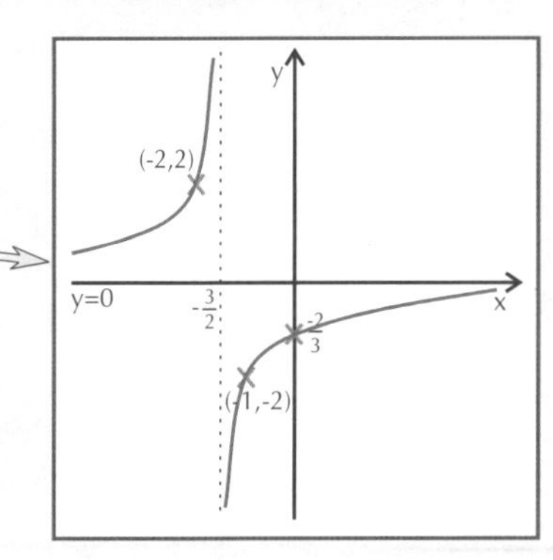

This is the same basic shape as $y=\frac{1}{x}$, but the extra numbers mean the basic shape has been moved and pulled around a bit. See pages 14 and 15 for more info.

Practice Questions

1) **Sketch the following curves:** a) $y = x^2 - 4x + 3$, b) $y = x^{10}$, c) $y = x^{-15}$, d) $y=\frac{3}{2x-5}$

2) **Sample Exam question:**

Find the asymptote parallel to the y-axis for the curve $y=\frac{6}{3x-2}$.
Hence sketch the curve, showing what happens as $x\to\infty$ and $x\to-\infty$. [4 marks]

Getting closer and closer but never quite arriving — like my train home on a Friday...

Find what happens to the graph at the axes, asymptotes, and as x gets really big. That's often enough to sketch the curve.

The Modulus Function

Skip these two pages if you're doing OCR B.

The modulus function makes graphs do weirdy things. Get your head round why this is.

The **Modulus** Function '**Bounces**' a Curve Off the **x-Axis**

Modulus signs (little vertical bars) make negative things positive, so for example, |-2| = 2.
Positive things aren't changed, so |2| = 2.

Example: Sketch the graphs of a) $y=|2x-4|$, b) $y=|x^2-2x-3|$, c) $y=-|3x|$, d) $y=-3|x|+2$.

Parts a) and b) are easy — the modulus bars are around everything on the right-hand side, so you know the graph can never go below the x-axis (as y can never be negative).

a) $y=|2x-4|$ — the easiest thing is to ignore the modulus signs at first, so draw the graph of $y = 2x - 4$...

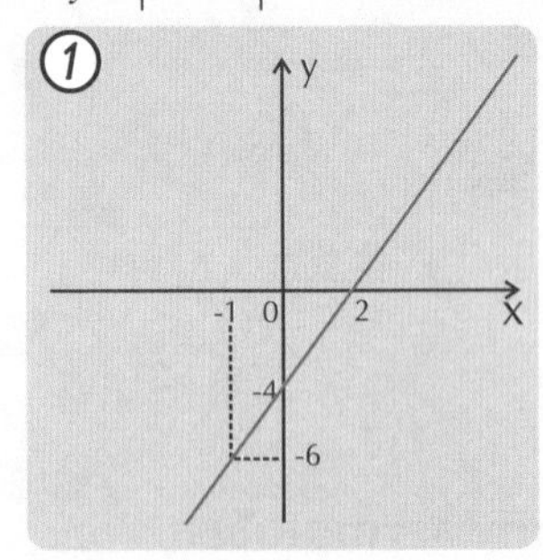

But |2x – 4| can't be negative, so the bit below the x-axis gets made positive...

...and the graph looks like this:

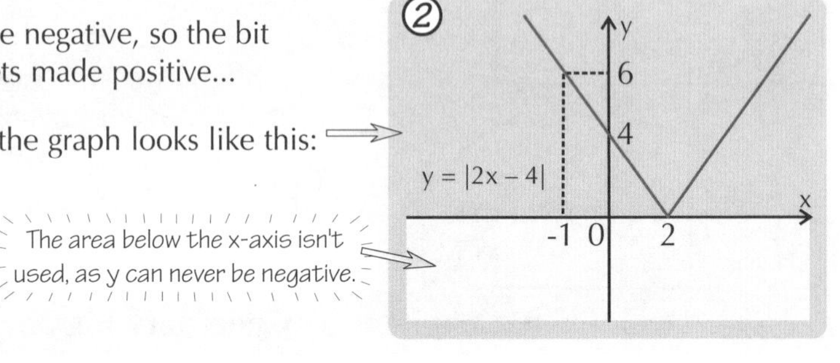

b) $y=|x^2-2x-3|$ — same again, draw the graph of $y=x^2-2x-3$ (or $y=(x-3)(x+1)$) first...

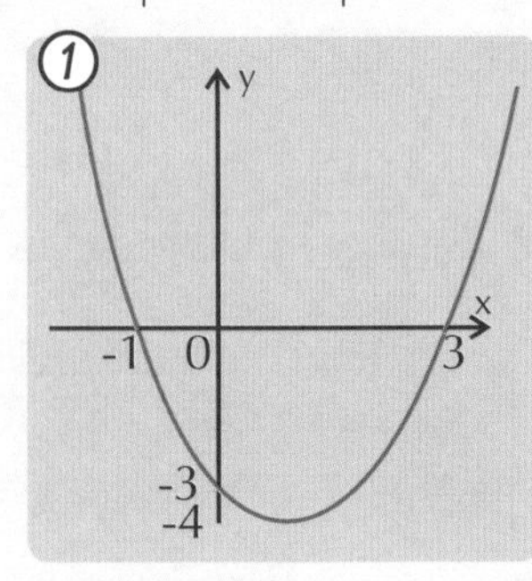

Again, y can't be negative, so make the bit below the x-axis positive...

...to get this:

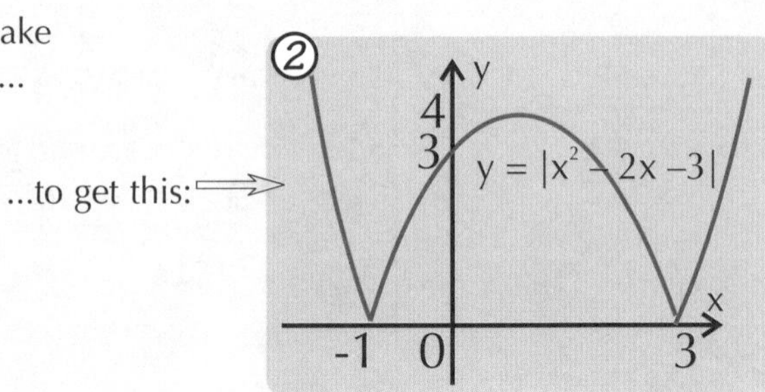

c) This time the function has a minus sign in front of the modulus — so now everything is made negative.

$y=-|3x|$ — reflect the graph of $y=|3x|$ in the x-axis:

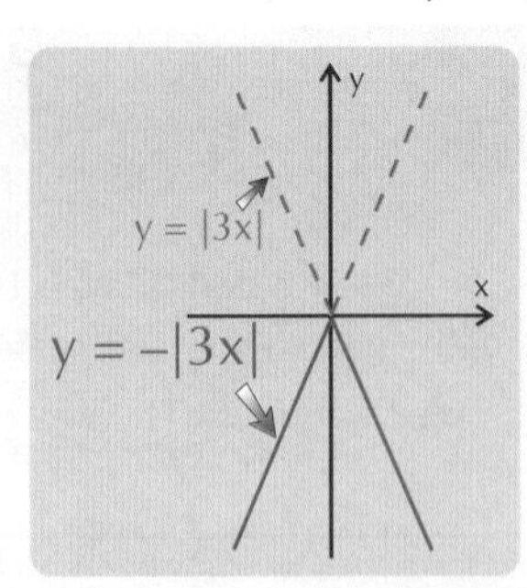

d) Now the modulus sign is around just one part of the expression for y, so you have to be even more careful...

$y=-3|x|+2$ — Do this in two parts.
First draw the graph for positive x, and then for negative x:

If x is positive, then |x| = x
— so for positive x, y = -3x + 2.

If x is negative, then |x| = -x
— so for negative x, y = -3(-x) + 2, i.e. y = 3x + 2

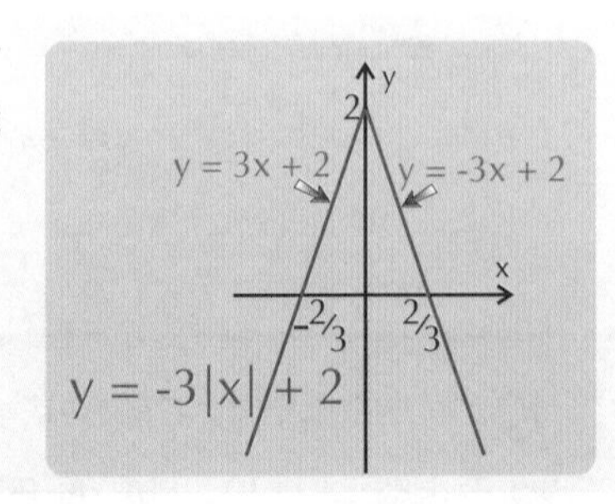

The Modulus Function

Modulus Questions need to be done in Two Parts

First you need to assume that the thing inside the modulus signs is positive, and then that it's negative.

Example: Let $f(x) = |2x + a|$, where a is positive. Find where the graph of $y = f(x) - a$ cuts the x-axis.

You're looking for where $|2x + a| - a = 0$. Do this in two stages...

① Assume $2x + a$ is positive: $(2x+a)-a=0$
$2x=0$, i.e. $x=0$

② Assume $2x + a$ is negative: $-(2x+a)-a=0$
$-2x-2a=0$,
i.e. $x=-a$

If '2x + a' is negative, then $|2x + a| = -(2x + a)$.

So the graph of $y = f(x) - a$ cuts the x-axis at (0, 0) and (-a, 0).

The Modulus Sign also crops up in Inequalities

Example: Solve the inequalities: a) $2|x| > 6$, b) $|x + 3| < 7$, c) $2|x| > |x + 3|$

The modulus of a number is just its 'size', i.e. its distance from O on a number line.

a) This kind is easy:

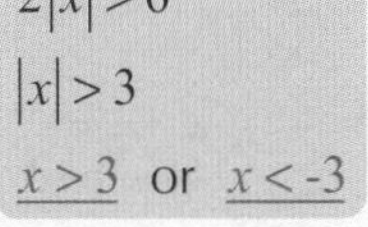

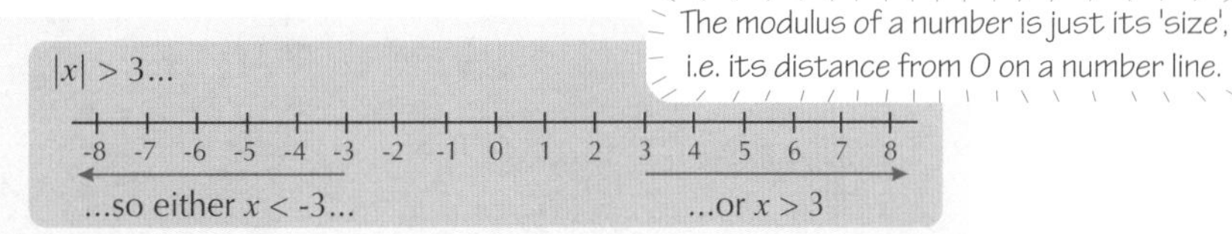

b) This kind isn't too bad either: $|x+3|<7$
$-7<x+3<7$
$-10<x<4$

Think of a number line again — if the modulus of something is less than 7, then it must be between -7 and 7.

c) This kind is a pain — although a sketch can make things slightly less confusing:

① Sketch both $y=2|x|$ and $y=|x+3|$ on the same set of axes.
You want the bits where $y=2|x|$ is above $y=|x+3|$.

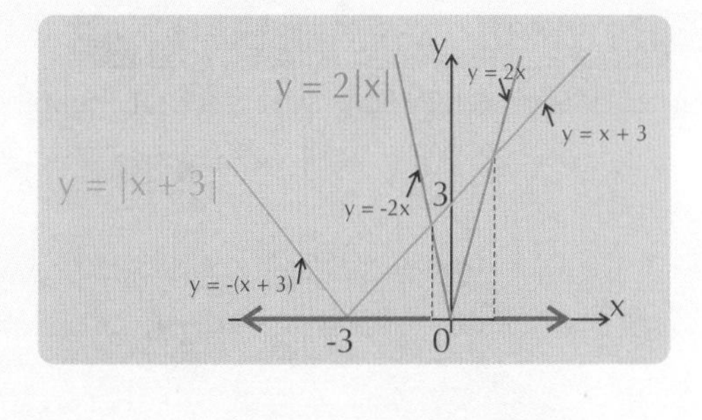

② Find where the two lines cross. From the graph you can see you need the point where the line y = x + 3 crosses y = 2x and the point where the line y = x + 3 crosses y = -2x.

So solve (i) $x+3=2x$, i.e. $x=3$, and (ii) $x+3=-2x$, i.e. $x=-1$.

③ Write the solution using inequality signs. From the graph, you can see this is where $x<-1$ or $x>3$.

Practice Questions

1) **Sketch the following graphs: a) $y = |3x - 6|$, b) $y = 3|x| - 6$**

2) **Solve the following inequalities: a) $4|x| < 12$, b) $|x - 4| > 3$, c) $|x| < |3x - 2|$**

3) **Sample Exam question:**

a)	Sketch the graph of $y = -|a - x|$, where a is a constant greater than 1.	[2 marks]
b)	On the same set of axes, sketch the graph of $y = x^2 - 2ax$.	[2 marks]
c)	If $a = 2$, find the values of x where the two curves intersect.	[5 marks]

Some of these graphs are a bit kinky — in the sense that they have kinks in...

Modulus questions can be confusing, which is why it's best to get quite a lot of practice at them. The thing to remember is that anything in the modulus sign could be positive, but it could be negative, and you have to allow for both possibilities. This often means doing the question twice, which is annoying. But that number-line trick is a good one and worth learning.

The Remainder and Factor Theorems

Skip these two pages if you're doing Edexcel, AQA A or AQA B.

Algebraic division is one of those things that you have to learn when you do AS maths. You'll probably never use it again once you've done your Exam, but hey ho... such is life.

Do *Polynomial Division* by means of *Subtraction*

$$(2x^3 - 3x^2 - 3x + 7) \div (x - 2) = ?$$

The trick with this is to see how many times you can subtract $(x-2)$ from $2x^3 - 3x^2 - 3x + 7$.
The idea is to keep subtracting lumps of $(x-2)$ until you've got rid of all the powers of x.

Do the subtracting in *Stages*

At each stage, always try to get rid of the highest power of x.
Then start again with whatever you've got left.

1. Start with $2x^3 - 3x^2 - 3x + 7$, and subtract $2x^2$ lots of $(x - 2)$ to get rid of the x^3 term.

$$(2x^3 - 3x^2 - 3x + 7) - 2x^2(x-2)$$
$$(2x^3 - 3x^2 - 3x + 7) - 2x^3 + 4x^2$$
$$= x^2 - 3x + 7$$

$2x^3 \div x = 2x^2$

This is what's left — so now you have to get rid of the x^2 term.

2. Now start again with $x^2 - 3x + 7$. The highest power of x is the x^2 term. So subtract x lots of $(x - 2)$ to get rid of that.

$$(x^2 - 3x + 7) - x(x-2)$$
$$(x^2 - 3x + 7) - x^2 + 2x$$
$$= -x + 7$$

Now start again with this — and get rid of the x term.

3. All that's left now is $-x + 7$. Get rid of the $-x$ by subtracting -1 times $(x - 2)$.

$$(-x+7) - (-1(x-2))$$
$$(-x+7) + x - 2$$
$$= 5$$

There are no more powers of x to get rid of — so stop here.

The remainder's 5.

Interpreting the results...

Time to work out exactly what all that meant.

Started with: $2x^3 - 3x^2 - 3x + 7$

Subtracted: $2x^2(x-2) + x(x-2) - 1(x-2)$

$= (x-2)(2x^2 + x - 1)$

Remainder: $= 5$

So... $2x^3 - 3x^2 - 3x + 7 = (x-2)(2x^2 + x - 1) + 5$

or... $\dfrac{2x^3 - 3x^2 - 3x + 7}{(x-2)} = 2x^2 + x - 1$ with remainder 5.

Algebraic Division

$$(ax^3 + bx^2 + cx + d) \div (x - k) = ?$$

1) SUBTRACT a multiple of $(x - k)$ to get rid of the highest power of x.
2) REPEAT step 1 until you've got rid of all the powers of x.
3) WORK OUT how many lumps of $(x - k)$, you've subtracted, and the REMAINDER.

The *Remainder Theorem* is an easy way to work out *Remainders*

When you divide $f(x)$ by $(x - a)$, the remainder is $f(a)$.

So in the example above, you could have worked out the remainder dead easily.

1) $f(x) = 2x^3 - 3x^2 - 3x + 7$.
2) You're dividing by $(x - 2)$, so $a = 2$.
3) So the remainder must be $f(2) = (2 \times 8) - (3 \times 4) - (3 \times 2) + 7 = 5$.

Careful now... when you're dividing by something like $(x + 7)$, a is negative — so here, $a = -7$.

The Remainder and Factor Theorems

The Remainder Theorem's easy, but it's not as useful as its little brother — the Factor Theorem.

The *Factor Theorem* is just the Remainder Theorem with a *Zero Remainder*

If you get a remainder of zero when you divide f(x) by (x – a), then (x – a) must be a factor. That's the Factor Theorem.

> The Factor Theorem:
> If f(x) is a polynomial, and f(a) = 0, then (x – a) is a factor of f(x).
> In other words: If you know the roots, you also know the factors — and vice versa.

Example: Show that (2x + 1) is a factor of $f(x) = 2x^3 - 3x^2 + 4x + 3$

The question's giving you a big hint here. Notice that $2x + 1 = 0$ when $x = -\frac{1}{2}$. So plug this value of x into $f(x)$.
If you show that $f(-\frac{1}{2}) = 0$, then the factor theorem says that $(x + \frac{1}{2})$ is a factor — which means that $2 \times (x + \frac{1}{2}) = (2x + 1)$ is also a factor.

$f(x) = 2x^3 - 3x^2 + 4x + 3$ and so $f\left(-\frac{1}{2}\right) = 2\times\left(-\frac{1}{8}\right) - 3\times\frac{1}{4} + 4\times\left(-\frac{1}{2}\right) + 3 = 0$

So, by the factor theorem, $(x + \frac{1}{2})$ is a factor of $f(x)$, and so $(2x + 1)$ is also a factor.

(x – 1) is a Factor if the coefficients *Add Up To 0*

This is a useful thing to remember.
It works for all polynomials — no exceptions.
It could save a fair whack of time in the exam.

Example: Factorise the polynomial $f(x) = 6x^2 - 7x + 1$

The coefficients (6, –7 and 1) add up to 0. That means $f(1) = 0$.
(And that applies to any polynomial at all... always.)

So by the factor theorem, if $f(1) = 0$, $(x - 1)$ is a factor. Easy.

Then just factorise it like any quadratic to get this:

$$f(x) = 6x^2 - 7x + 1 = (6x - 1)(x - 1)$$

Practice Questions

1) *Write the following functions f(x) in the form f(x) = (x + 2)g(x) + remainder (where g(x) is a quadratic):*
a) $f(x) = 3x^3 - 4x^2 - 5x - 6$, b) $f(x) = x^3 + 2x^2 - 3x + 4$

2) *Find the remainder when the following are divided by: (i) (x + 1), (ii) (x – 1)*
a) $f(x) = 6x^3 - x^2 - 3x - 12$, b) $f(x) = x^4 + 2x^3 - x^2 + 3x + 4$

3) *Sample Exam question:*

Find the values of c and d so that $2x^4 + 3x^3 + 5x^2 + cx + d$ is exactly divisible by $(x - 2)(x + 3)$. [6 marks]

The Remainder Theorem is just a factor life...

The Remainder Theorem and Factor Theorem are easy. To be honest, the Remainder Theorem's not a whole heap of use except for checking that you've done your algebraic division right. The Factor Theorem is worth practising a fair bit, though, since it can make factorising quadratics and cubics a whole lot easier and quicker.

Graph Transformations

Skip these two pages if you're doing OCR A or AQA B.

Suppose you start with any old function f(x).
Then you can transform (change) it in three ways — by translating it, stretching or reflecting it.

y = f(x)

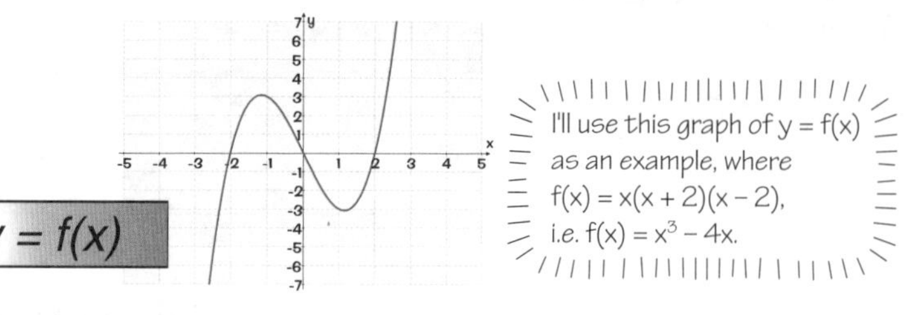

Translations are caused by *Adding* things

y = f(x)+a

Adding a number to the whole function shifts the graph up or down.

1) If $a > 0$, the graph goes upwards.
2) If $a < 0$, the graph goes downwards.

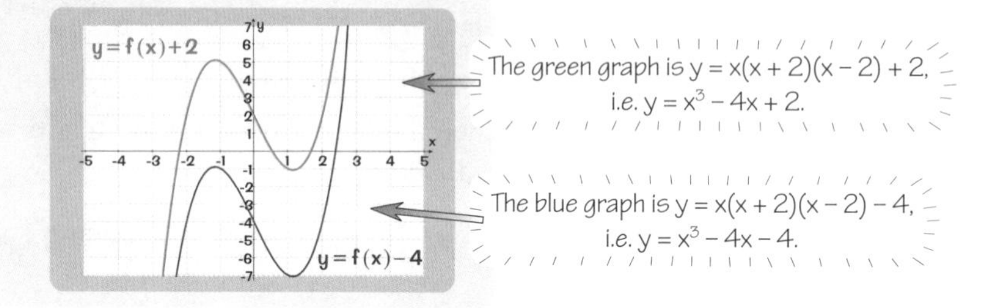

y = f(x+a)

Writing 'x + a' instead of 'x' means the graph moves sideways.

1) If $a > 0$, the graph goes to the left.
2) If $a < 0$, the graph goes to the right.

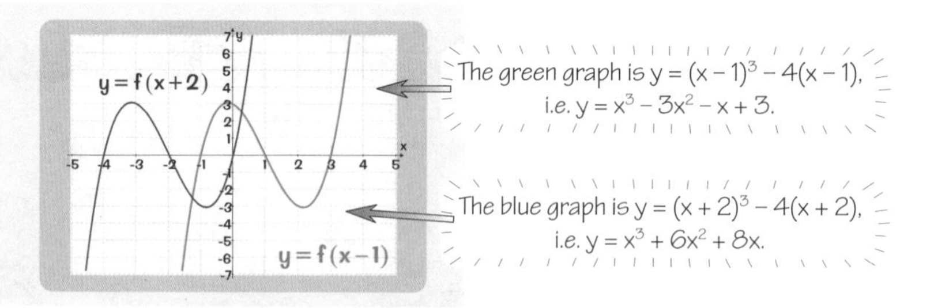

Stretches and *Reflections* are caused by *Multiplying* things

y = af(x)

Multiplying the whole function stretches, squeezes or reflects the graph vertically.

1) Negative values of 'a' reflect the basic shape in the x-axis.
2) If $a > 1$ or $a < -1$ (i.e. $|a| > 1$) the graph is stretched vertically.
3) If $-1 < a < 1$ (i.e. $|a| < 1$) the graph is squashed vertically.

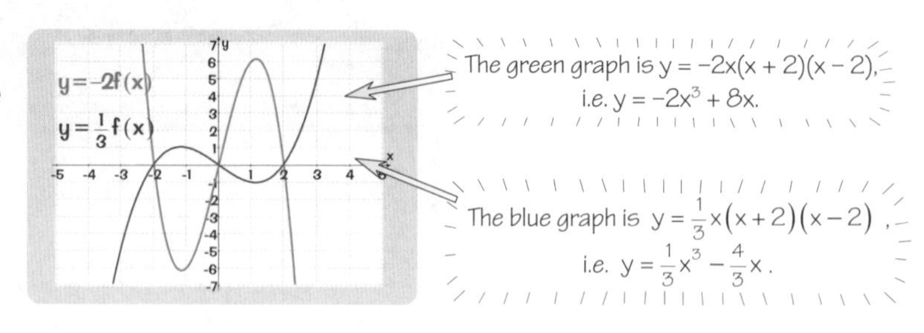

y = f(ax)

Writing 'ax' instead of 'x' stretches, squeezes or reflects the graph horizontally.

1) Negative values of 'a' reflect the basic shape in the y-axis.
2) If $a > 1$ or $a < -1$ (i.e. if $|a| > 1$) the graph is squashed horizontally.
3) If $-1 < a < 1$ (i.e. if $|a| < 1$) the graph is stretched horizontally.

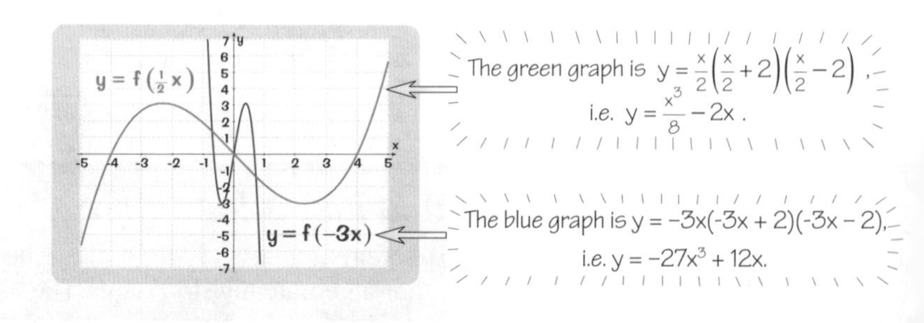

Graph Transformations

The Modulus function can be used in Transformations too

Example: If $f(x) = (x + 1)(x - 2)(x - 3)$, sketch: (i) $y = f(x)$, (ii) $y = |f(x)|$, (iii) $y = f(|x|)$.

(i) A normal cubic with zeros at -1, 2 and 3.

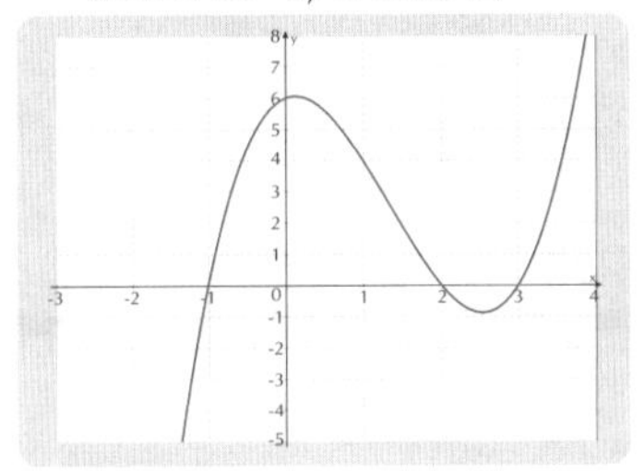

(ii) Everything's positive...

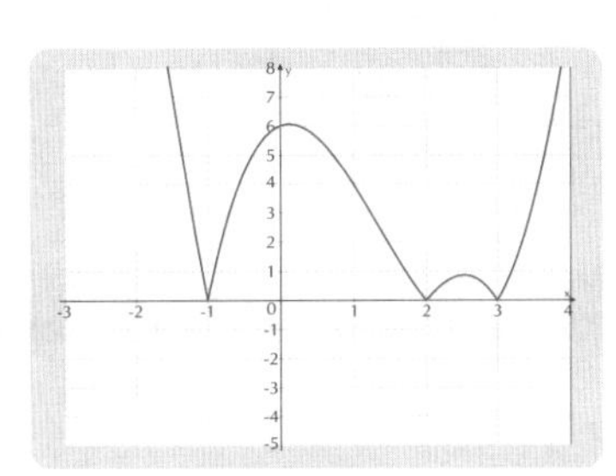

(ii) Everything on the left of the y-axis is the same as on the right.

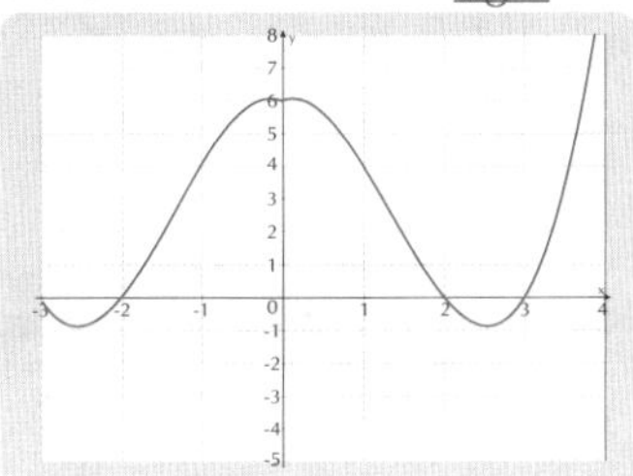

It's exactly the same for Sin, Cos and Tan graphs

If you're doing OCR B, you need to be able to do all this stuff for trigonometric functions as well — but it's no different.

Example: Sketch $y = 3\sin(2x - 60°)$ for $0° \leq x \leq 180°$.

There are 3 changes to make to a common-or-garden sine graph like this one.
It isn't hard, but there's a wee catch you need to be a bit wary of...

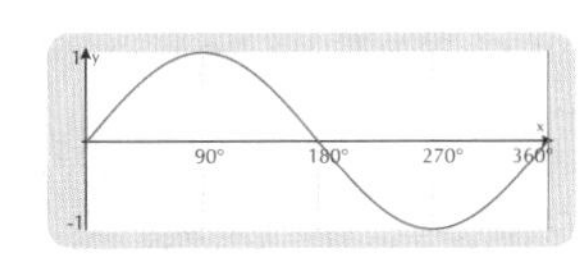

Deal with the easy stuff first...

$3\sin(2x - 60°)$ — the '3' means you need to stretch it by a factor of 3 vertically.
$3\sin(2x - 60°)$ — the '2' means you need to squash it by a factor of 2 horizontally.

So this gives you the graph of $y = 3\sin 2x$ (for $0° \leq x \leq 360°$) — so far, so good...

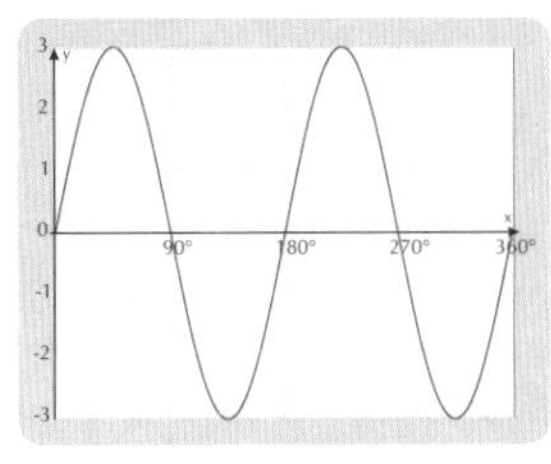

Now the tricky bit — how much to shift it across.
It's tempting to just whack it across 60° to the right — but think about it...

$3\sin 2x = 0$ when $2x = 0$, i.e. when $x = 0$.
This means $3\sin(2x - 60°) = 0$ when $2x - 60° = 0$
— i.e. when $x = 30°$.
So when you draw the graph, it looks as though it's just been shifted 30° to the right:

Think of it as $y = 3\sin(2(x - 30°))$.

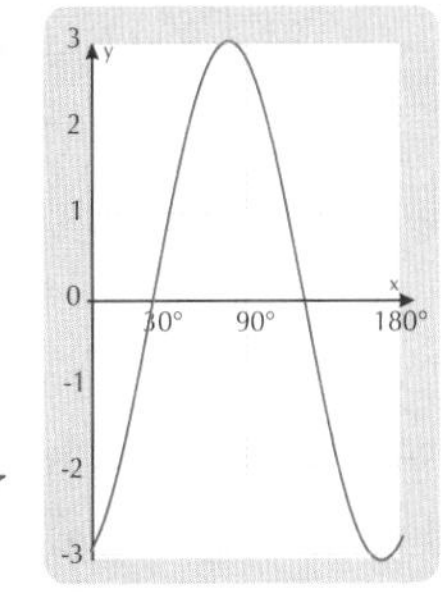

Practice Questions

1) If $f(x) = (x - 3)(x - 2)(x + 1)$, sketch a) $y = f(2x)$ and b) $y = -f(|x|)$.

2) Sample exam question:

a) Sketch the graph of $y = 1 - (x - 2)^2$, marking carefully the points where the curve meets the coordinate axes. [3 marks]

b) Using the same axes, sketch the graph of $y = 1.5x - 2$. [1 mark]

c) Prove that the x-coordinates of the points of intersection of the two graphs satisfy the equation $2x^2 - 5x + 2 = 0$. [3 marks]

d) Solve this equation to find the coordinates of the points of intersection of the two graphs. [3 marks]

Stretching, squeezing and translating — not bad for two pages of maths...

With this graph-sketching and graph-transforming business, it's always a good idea to check one or two points once you've done your sketch. Stick in a couple of values of x, and if the y-values aren't what you think they should be, you'll need to think again. Practise this stuff — but if you make any mistakes, work out what went wrong before trying another question.

Sequences

Skip these two pages if you're doing AQA B.

A sequence is a list of numbers that follow a certain pattern. Sequences can be finite or infinite (infinity — oooh), and they're usually generated in one of two ways. And guess what? You have to know everything about them.

A Sequence can be defined by its nth Term

You almost definitely covered this stuff at GCSE, so no excuses for mucking it up.

The point of all this is to show how you can work out any value (the nth term) from its position in the sequence (n).

Example: Find the nth term of the sequence 5, 8, 11, 14, 17, ...

1st	2nd	3rd	4th	5th
5	8	11	14	17

+3 +3 +3 +3

Each term is 3 more than the one before it. That means that you need to start by multiplying n by 3.

Take the first term (where n = 1). If you multiply n by 3, you still have to add 2 to get 5.

The same goes for n = 2. To get 8 you need to multiply n by 3, then add 2.
Every term in the sequence is worked out exactly the same way.

So n^{th} term is $3n + 2$

You can define a sequence by a Recurrence Relation too

Don't be put off by the fancy name — recurrence relations are pretty easy really.

The main thing to remember is:
a_k just means the kth term of the sequence

The next term in the sequence is a_{k+1}. You need to describe how to work out a_{k+1} if you're given a_k.

Example: Find the recurrence relation of the sequence 5, 8, 11, 14, 17, ...

From the example above, you know that each term equals the one before it, plus 3.

This is written like this: $\mathbf{a_{k+1} = a_k + 3}$

So, if k = 5, $a_k = a_5$ which stands for the 5th term, and $a_{k+1} = a_6$ which stands for the 6th term.

In everyday language, $a_{k+1} = a_k + 3$ means that the sixth term equals the fifth term plus 3.

BUT $a_{k+1} = a_k + 3$ on its own isn't enough to describe 5, 8, 11, 14, 17,...
For example, the sequence 87, 90, 93, 96, 99, ... also has each term being 3 more than the one before.

The recurrence relation needs to be more specific, so you've got to give one term in the sequence.
You almost always give the first value, a_1.

Putting all of this together gives 5, 8, 11, 14, 17,... as $a_{k+1} = a_k + 3, \ a_1 = 5$

Sequences

Some sequences involve **Multiplying**

You've done the easy 'adding' business. Now it gets really tough — multiplying. Are you sure you're ready for this...

Example: A sequence is defined by $a_{k+1} = 2a_k - 1$, $a_2 = 5$. List the first five terms.

OK, you're told the second term, $a_2 = 5$. Just plug that value into the equation, and carry on from there.

$a_3 = 2 \times 5 - 1 = 9$ ⟸ From the equation $a_k = a_2$ so $a_{k+1} = a_3$

$a_4 = 2 \times 9 - 1 = 17$ ⟸ Now use a_3 to find $a_{k+1} = a_4$ and so on...

$a_5 = 2 \times 17 - 1 = 33$

Now to find the first term, a_1:

$a_2 = 2a_1 - 1$ ⟸ Just make $a_k = a_1$

$5 = 2a_1 - 1$

$2a_1 = 6$

$a_1 = 3$

So the first five terms of the sequence are 3, 5, 9, 17, 33.

Some Sequences have a **Certain Number** of terms — others go on **Forever**

Some sequences are only defined for a certain number of terms.

It's the $1 \le k \le 20$ bit that tells you it's finite.

For example, $a_{k+1} = a_k + 3$, $a_1 = 1$, $1 \le k \le 20$ will be 1, 4, 7, 10, ..., 58 and will contain 20 terms.
This is a finite sequence.

Other sequences don't have a specified number of terms and could go on forever.

For example, $u_{k+1} = u_k + 2$, $u_1 = 5$, will be 5, 7, 9, 11, 13, ... and won't have a final term.
This is an infinite sequence.

While others are periodic, and just revisit the same values over and over again.

For example, $u_k = u_{k-3}$, $u_1 = 1$, $u_2 = 4$, $u_3 = 2$, will be 1, 4, 2, 1, 4, 2, 1, 4, 2,...
This is a periodic sequence with period 3.

Practice Questions

1) A sequence has an nth term of $n^2 + 3$. Find a) the first four terms, and b) the 20th term.

2) A sequence is defined by $a_{k+1} = 3a_k - 2$. If $a_1 = 4$, find a_2, a_3 and a_4.

3) A sequence of integers, $u_1, u_2, u_3, \ldots$ is given by $u_{n+1} = u_n + 5$, $u_1 = 3$. Write an expression for the nth term of this sequence.

Like maths teachers, sequences can go on and on and on and on...

If you know the formula for the nth term, you can work out any term using a single formula, so it's kind of easy. If you only know a recurrence relation, then you can only work out the next term. So if you want the 20th term, and you only know the first one, then you have to use the recurrence relation 19 times. (So it'd be quicker to work out a formula really.)

Arithmetic Progressions

Skip these two pages if you're doing Edexcel or AQA B.

Right, you've got basic sequences tucked under your belt now — time to step it up a notch (sounds painful). When the terms of a sequence progress by adding a fixed amount each time, this is called an arithmetic progression.

*It's all about **Finding the nth Term***

The first term of a sequence is given the symbol **a**. The amount you add each time is called the common difference, called **d**. The position of any term in the sequence is called **n**.

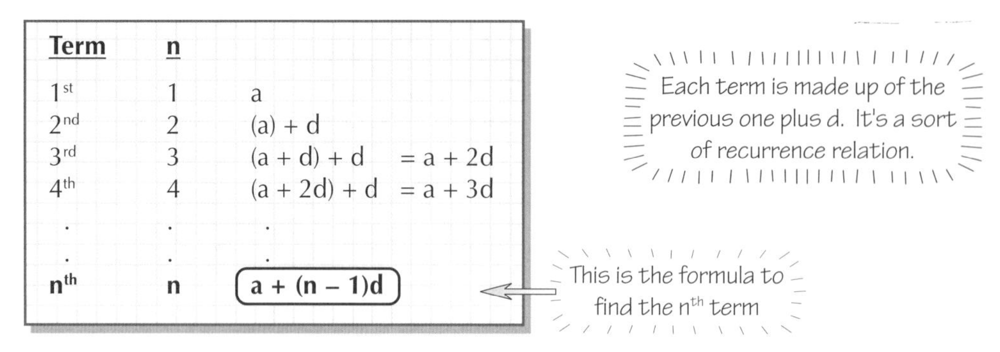

Term	n		
1st	1	a	
2nd	2	$(a) + d$	
3rd	3	$(a + d) + d$	$= a + 2d$
4th	4	$(a + 2d) + d$	$= a + 3d$
.	.	.	
.	.	.	
n^{th}	n	$a + (n - 1)d$	

Example: Find the 20th term of the arithmetic progression 2, 5, 8, 11,... and find the formula for the nth term.

Here $a = 2$ and $d = 3$ — *To get d, just find the difference between two terms next to each other — e.g. 11 – 8 = 3*

So 20th term $= a + (20 - 1)d$
$= 2 + 19 \times 3$
$= 59$

The general term is the nth term, i.e.
$a + (n - 1)d$
$= 2 + (n - 1)3$
$= 3n - 1$

*A **Sequence** becomes a **Series** when you **Add the Terms** to **Find the Total***

S_n is the total of the first n terms of the arithmetic progression:

$$S_n = a + (a + d) + (a + 2d) + (a + 3d) + \ldots + (a + (n - 1)d)$$

There's a really neat version of the same formula too:

$$S_n = n \times \frac{(a + l)}{2}$$

The l stands for the last value in the progression. You work it out as $l = a + (n - 1)d$

Nobody likes formulas, so think of it as the average of the first and last terms multiplied by the number of terms.

Example: Find the sum of the series with first term 3, last term 87 and common difference 4.

Here you know a, d and l, but you don't know n yet.

Use the information about the last value, l: $a + (n - 1)d = 87$

Then plug in the other values: $3 + 4(n - 1) = 87$
$4n - 4 = 84$
$4n = 88$
$n = 22$

So $S_{22} = 22 \times \frac{(3+87)}{2}$ $\quad S_{22} = 990$

Arithmetic Progressions

They Won't always give you the Last Term...

...but don't panic — there's a formula to use when the last term is unknown. But you knew I'd say that, didn't you?

You know $l = a + (n-1)d$ and $S_n = n\frac{(a+l)}{2}$.

Plug l into S_n and rearrange to get the formula in the box:

$$S_n = \frac{n}{2}[2a + (n-1)d]$$

Example: For the sequence -5, -2, 1, 4, 7, ... find the sum of the first 20 terms.

So $a = -5$ and $d = 3$. The question says $n = 20$ too.

$$S_{20} = \frac{20}{2}[2\times -5 + (20-1)\times 3]$$
$$= 10[-10 + 19\times 3]$$
$$S_{20} = 470$$

There's Another way of Writing Series, too

So far, the letter S has been used for the sum. The Greeks did a lot of work on this — their capital letter for S is sigma, or Σ. This is used today, together with the general term, to mean the sum of the series.

Example: Find $\sum_{n=1}^{15}(2n+3)$

...and ending with n=15

Starting with n=1...

This means you have to find the sum of the first 15 terms of the series with n^{th} term $2n + 3$.

The first term ($n = 1$) is 5, the second term ($n = 2$) is 7, the third is 9, ... and the last term ($n = 15$) is 33. In other words, you need to find 5 + 7 + 9 + ... + 33. This gives $a = 5$, $d = 2$, $n = 15$ and $l = 33$.

You know all of a, d, n and l, so you can use either formula:

$$S_n = n\frac{(a+l)}{2}$$
$$S_{15} = 15\frac{(5+33)}{2}$$
$$S_{15} = 15 \times 19$$
$$S_{15} = 285$$

It makes no difference which method you use.

$$S_n = \frac{n}{2}[2a + (n-1)d]$$
$$S_{15} = \frac{15}{2}[2\times 5 + 14\times 2]$$
$$S_{15} = \frac{15}{2}[10 + 28]$$
$$S_{15} = 285$$

Practice Questions

1) ***Find the sum of the series that begins with 5, 8, ... and ends with 65.***

2) ***A series has first term 7 and 5th term 23.***
Find a) the common difference, b) the 15th term, and c) the sum of the first ten terms.

3) ***A series has seventh term 36 and tenth term 30. Find the sum of the first five terms and the nth term.***

4) ***Find*** $\sum_{n=1}^{20}(3n-1)$

5) ***Find*** $\sum_{n=1}^{10}(48-5n)$

This sigma notation is all Greek to me... *(Ho ho ho)*

A sequence is just a list of numbers (with commas between them) — a series on the other hand is when you add all the terms together. It doesn't sound like a big difference, but mathematicians get all hot under the collar when you get the two mixed up. Remember that BlackADDer was a great TV series — not a TV sequence. (Sounds daft, but I bet you remember it now.)

Arithmetic Progressions

Skip these two pages if you're doing Edexcel or AQA B.

Use **Arithmetic Progressions** to add up the **First n Whole Numbers**

The sum of the first n natural numbers looks like this: $S_n = 1 + 2 + 3 + \ldots + (n-2) + (n-1) + n$

So $a = 1$, $l = n$ and also $n = n$. Now just plug those values into the formula:

$$S_n = n \times \frac{(a+l)}{2} \Longrightarrow S_n = \frac{1}{2}n(n+1)$$

Natural numbers are just positive whole numbers.

Example: Add up all the whole numbers from 1 to 100.

Sounds pretty hard, but all you have to do is stick it into the formula:

$S_{100} = \frac{1}{2} \times 100 \times 101$. So $S_{100} = 5050$

Series **Don't** have to start with **n = 1**

Instead of adding up numbers from 1 to 100, what do you do if you want to add the natural numbers from 50 to 100? This just means the sum from 1 to 100, but without the first 49 whole numbers.

You can write this as

$$\sum_{n=50}^{100} n = \sum_{n=1}^{100} n - \sum_{n=1}^{49} n$$
$$= 5050 - \frac{49 \times 50}{2}$$
$$= 5050 - 1225$$
$$= 3825$$

From the example above.

Using $Sn = \frac{n(n+1)}{2}$

Subtract any **Series** if it **Doesn't** Start at **n = 1**

Example: Find $\sum_{n=7}^{20}(4n-1)$

$$\sum_{n=7}^{20}(4n-1) = \sum_{n=1}^{20}(4n-1) - \sum_{n=1}^{6}(4n-1)$$
$$= \frac{20(3+79)}{2} - \frac{6(3+23)}{2}$$
$$= 820 - 78$$
$$= 742$$

Using $S_n = n\frac{(a+l)}{2}$

NB: $\sum_{n=1}^{20}(4n-1)$ and $\sum_{n=1}^{6}(4n-1)$ could both have been worked out a different way:

$$\sum_{n=1}^{6}(4n-1) = \sum_{n=1}^{6}4n - \sum_{n=1}^{6}1$$
$$= 4\sum_{n=1}^{6}n - \sum_{n=1}^{6}1$$
$$= 4\left(\frac{6 \times 7}{2}\right) - 6$$
$$= 78$$

Because $\sum_{n=1}^{6}4n = 4\sum_{n=1}^{6}n$

$\sum_{n=1}^{6}1 = 1+1+1+1+1+1 = 6$

1) Sample exam question:

a) An arithmetic series has first term a and common difference d.
Prove that the sum of the first n terms, S_n, is given by the formula $S_n = \frac{n}{2}[2a + (n-1)d]$. [4 marks]

b) Evaluate $\sum_{n=9}^{32}(2n-5)$. [4 marks]

Geometric Progressions

So arithmetic progressions mean you add a number to get the next term.
Geometric progressions are a bit different — you multiply by a number to get the next term.

Geometric Progressions Multiply by a Constant each time

Geometric progressions work like this: the next term in the sequence is obtained by multiplying the previous one by a constant value. Couldn't be easier.

$u_1 = a \qquad\qquad\qquad = a$

$u_2 = a \times r \qquad\qquad = ar$

$u_3 = a \times r \times r \qquad = ar^2$

$u_4 = a \times r \times r \times r = ar^3$

The first term (u_1) is called 'a'.

The number you multiply by each time is called 'the common ratio', symbolised by 'r'.

Here's the formula describing any term in the geometric progression: $u_n = ar^{n-1}$

Example: There is a chessboard with a 1p piece on the first square, 2p on the second square, 4p on the third, 8p on the forth and so on until the board is full. Calculate how much money is on the board.

This is a geometric progression, where you get the next term in the sequence by multiplying the previous one by 2.
So a = 1 (because you start with 1p on the first square) and r = 2.

So $u_1=1, u_2=2, u_3=4, u_4=8, \ldots$

You often have to work out the Sum of the Terms

Just like before, S_n stands for the sum of the first n terms.
In the example above, you're told to work out S_{64} (because there are 64 squares on a chessboard).

To find the sum of a G.P. you use two series and subtract.

For a G.P.: $S_n = a + ar + ar^2 + ar^3 + \ldots + ar^{n-1}$

Multiplying by r gives: $rS_n = ar + ar^2 + ar^3 + \ldots + ar^{n-2} + ar^{n-1} + ar^n$

Subtracting gives: $S_n - rS_n = a - ar^n$

Factorising: $(1-r)S_n = a(1-r^n) \Rightarrow S_n = \dfrac{a(1-r^n)}{1-r}$

If the series were subtracted the other way around you'd get $S_n = \dfrac{a(r^n-1)}{r-1}$. Both versions are correct.

So, back to the chessboard example: $a = 1, r = 2, n = 64 \qquad S_{64} = \dfrac{1(1-2^{64})}{1-2}$

$S_{64} = 1.84 \times 10^{19}$ pence or £1.84×10^{17}

The whole is more than the sum of the parts — hmm, not in maths, it ain't...

You really need to understand the difference between arithmetic and geometric progressions — it's not hard, but it needs to be fixed firmly in your head. There are only a few formulas for sequences and series (the nth term of a sequence, the sum of the first n terms of a series), but you need to learn them, since they won't be in the formula book they give you.

Geometric Progressions

Skip these two pages if you're doing Edexcel.

Geometric progressions can either *Grow* or *Shrink*

In the chessboard example, each term was bigger than the previous one, 1, 2, 4, 8, 16, ...
You can create a series where each term is less than the previous one by using a small value of r.

Example: If $a = 20$ and $r = \frac{1}{5}$, write down the first five terms of the sequence and the 20th term.

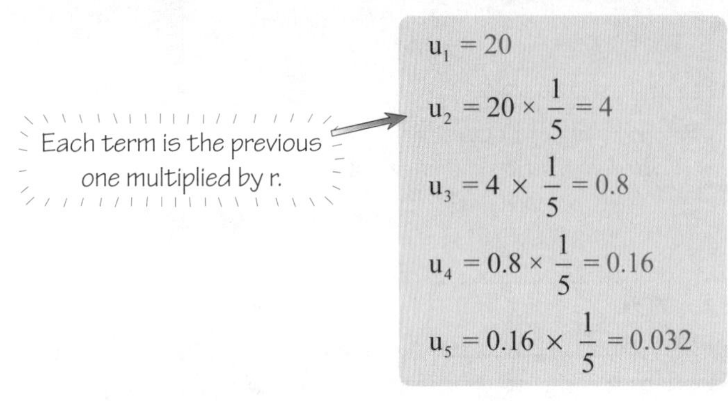

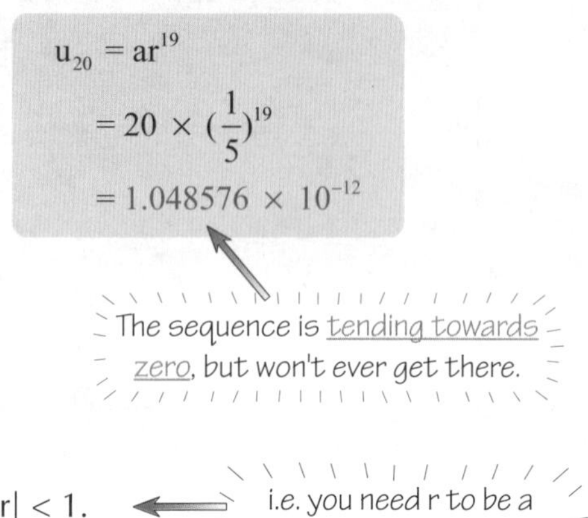

In general, for each term to be smaller than the one before, you need $|r| < 1$.
A sequence with $|r| < 1$ is called convergent.
Any other sequence (like the chessboard example on page 21) is called divergent.

i.e. you need r to be a fraction between -1 and 1 (see pages 10 and 11).

A *Convergent* series has a *Sum* to *Infinity*

In other words, if you just kept on with a convergent series, you'd get closer and closer to a certain number, but you'd never actually reach it.

If $|r| < 1$ and n is very, very big, then r^n will be very, very small — or to put it technically, $r^n \to 0$.
(Try working out $(\frac{1}{2})^{100}$ on your calculator if you don't believe me.)
This means $(1 - r^n)$ is really, really close to 1.

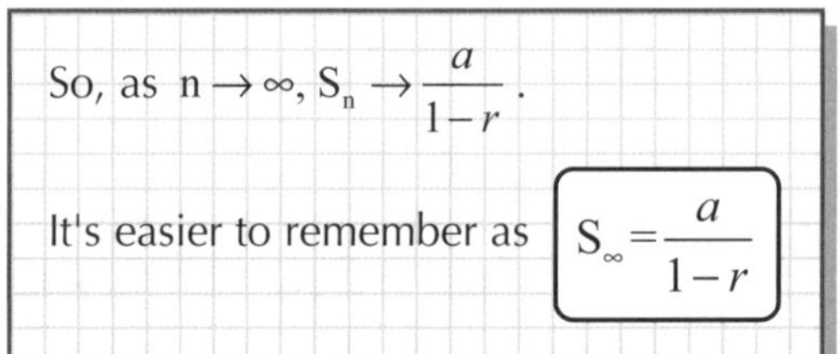

S_∞ just means 'sum to infinity'.

Example: If $a = 2$ and $r = \frac{1}{2}$, find the sum to infinity of the geometric series.

$u_1 = 2 \quad\Rightarrow\quad S_1 = 2$

$u_2 = 2 \times \frac{1}{2} = 1 \quad\Rightarrow\quad S_2 = 2 + 1 = 3$

$u_3 = 1 \times \frac{1}{2} = \frac{1}{2} \quad\Rightarrow\quad S_3 = 2 + 1 + \frac{1}{2} = 3\frac{1}{2}$

$u_4 = \frac{1}{2} \times \frac{1}{2} = \frac{1}{4} \quad\Rightarrow\quad S_4 = 2 + 1 + \frac{1}{2} + \frac{1}{4} = 3\frac{3}{4}$

$u_5 = \frac{1}{4} \times \frac{1}{2} = \frac{1}{8} \quad\Rightarrow\quad S_5 = 2 + 1 + \frac{1}{2} + \frac{1}{4} + \frac{1}{8} = 3\frac{7}{8}$

$u_6 = \frac{1}{8} \times \frac{1}{2} = \frac{1}{16} \quad\Rightarrow\quad S_6 = 2 + 1 + \frac{1}{2} + \frac{1}{4} + \frac{1}{8} + \frac{1}{16} = 3\frac{15}{16}$

These values are getting smaller each time.

These values are getting closer (converging) to 4.
So, the sum to infinity is 4.

You can show this graphically:
The line on the graph is getting closer and closer to 4, but it'll never actually get there.

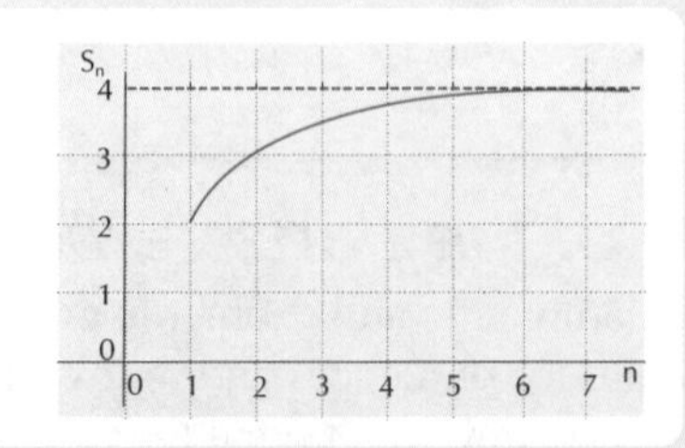

Of course, you could have saved yourself a lot of bother by using the sum to infinity formula:

$$S_\infty = \frac{a}{1-r} = \frac{2}{1-\frac{1}{2}} = 4$$

Geometric Progressions

A Divergent series Doesn't have a sum to infinity

Example:

If a = 2 and r = 2, find the sum to infinity of the series.

$u_1 = 2 \longrightarrow S_1 = 2$
$u_2 = 2 \times 2 = 4 \longrightarrow S_2 = 2 + 4 = 6$
$u_3 = 4 \times 2 = 8 \longrightarrow S_3 = 2 + 4 + 8 = 14$
$u_4 = 8 \times 2 = 16 \longrightarrow S_4 = 2 + 4 + 8 + 16 = 30$
$u_5 = 16 \times 2 = 32 \longrightarrow S_5 = 2 + 4 + 8 + 16 + 32 = 62$

As $n \to \infty$, $S_n \to \infty$ in a big way. So big, in fact, that eventually you can't work it out — so don't bother.

There is no sum to infinity for a divergent series.

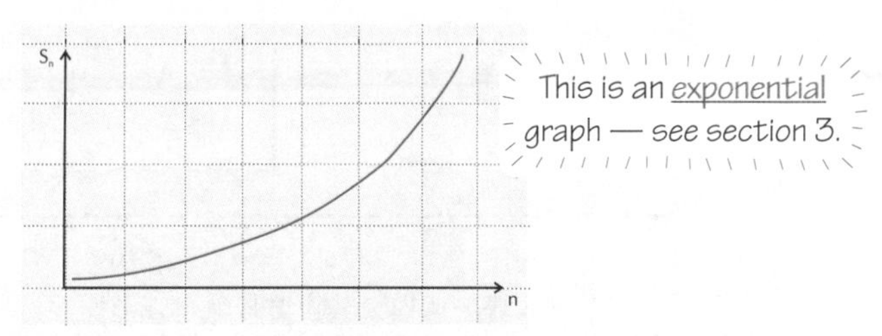

Example:

When a baby is born, £3000 is invested in an account with a fixed interest rate of 4% per year.
a) What will the account be worth at the start of the seventh year?
b) Will the account have doubled in value by the time the child reaches its 21st birthday?

a) $u_1 = a = 3000$

$u_2 = 3000 + (4\% \text{ of } 3000)$ ← This is the interest.
$= 3000 + (0.04 \times 3000)$
$= 3000\,(1 + 0.04)$
$= 3000 \times 1.04$ ← So, r = 1.04

$u_3 = u_2 \times 1.04$
$= (3000 \times 1.04) \times 1.04$
$= 3000 \times (1.04)^2$

$u_4 = 3000 \times (1.04)^3$
. . .
. . . ← I've missed out some steps here — check that you understand what's happened.

$u_7 = 3000 \times (1.04)^6$
$= £3795.96$ (to the nearest penny)

b) You need to know when $u_n > 6000$ ← double the original value.
From part a) you can tell that $u_n = 3000 \times (1.04)^{n-1}$
So $3000 \times (1.04)^{n-1} > 6000$
$(1.04)^{n-1} > 2$

To complete this you need to use logs (see section 3):

$\log(1.04)^{n-1} > \log 2$

$(n - 1)\log(1.04) > \log 2$

$$n - 1 > \frac{\log 2}{\log 1.04}$$

$n - 1 > 17.67$

$n > 18.67$ (to 2 d.p.)

So u_{19} (the amount at the start of the 19th year) will be more than double the original amount — plenty of time to buy a Porsche for the 21st birthday.

Practice Questions

1) For the sequence 2, -6, 18, ..., find the 10th term.

2) For the sequence 24, 12, 6, ..., find a) the common ratio, b) the seventh term, and c) the sum to infinity.

3) A G.P. and an A.P. both begin with 2, 6, ... Which term of the A.P. will be equal to the fifth term of the G.P.?

4) Sample exam question:

For the series with second term -2 and common ratio -½, find

a) the first term [3 marks]
b) the first seven terms [3 marks]
c) the sum of the first seven terms [3 marks]
d) the sum to infinity [3 marks]

So tell me — if my savings earn 4% per year, when will I be rich...

Now here's a funny thing — you can have a convergent geometric series if the common ratio is small enough. I find this odd — that I can keep adding things to a sum forever, but the sum never gets really really big.

Binomial Expansions

Skip these two pages if you're doing OCR A, OCR B, AQA A or AQA B.

If you're feeling a bit stressed, just take a couple of minutes to relax before trying to get your head round this page — it's a bit of a stinker in places. Have a cup of tea and think about something else for a couple of minutes. Ready...

Writing **Binomial Expansions** is all about **Spotting Patterns**

Doing binomial expansions just involves multiplying out the brackets. It would get nasty when you raise the brackets to higher powers — but once again I've got a cunning plan...

$$(1+x)^0 = 1$$
$$(1+x)^1 = 1+x$$
$$(1+x)^2 = 1+2x+x^2$$
$$(1+x)^3 = 1+3x+3x^2+x^3$$
$$(1+x)^4 = 1+4x+6x^2+4x^3+x^4$$

Anything to the power 0 is 1.

$$(1+x)^3 = (1+x)(1+x)^2$$
$$= (1+x)(1+2x+x^2)$$
$$= 1+2x+x^2+x+2x^2+x^3$$
$$= 1+3x+3x^2+x^3$$

A Frenchman named Pascal spotted the pattern in the coefficients and wrote them down in a triangle.
So it was called 'Pascal's Triangle' (imaginative, eh?).
The pattern's easy — each number is the sum of the two above it.

So, the next line will be: **1 5 10 10 5 1**
giving **$(1 + x)^5 = 1 + 5x + 10x^2 + 10x^3 + 5x^4 + x^5$.**

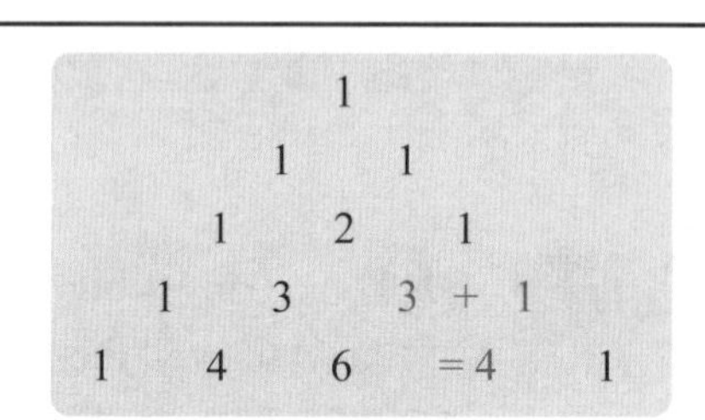

You **Don't** need to write out Pascal's Triangle for **Higher Powers**

There's a formula for the numbers in the triangle. The formula looks horrible (one of the worst in AS maths) so don't try to learn it letter by letter — look for the patterns in it instead. Here's an example:

Example: Expand $(1 + x)^{20}$, giving the first four terms only.

So you can use this formula for any power, the power is called n. In this example n = 20.

$$(1+x)^n = 1+\frac{n}{1}x+\frac{n(n-1)}{1\times2}x^2+\boxed{\frac{n(n-1)(n-2)}{1\times2\times3}x^3}+.............+x^n$$

Here's a closer look at the term in the black box:

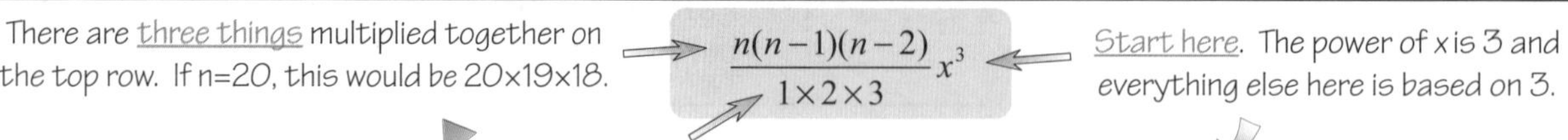

This means, if n = 20 and you were asked for 'the term in x^7' you should write $\frac{20\times19\times18\times17\times16\times15\times14}{1\times2\times3\times4\times5\times6\times7}x^7$.

This can be simplified to $\frac{20!}{7!13!}x^7$

$20\times19\times18\times17\times16\times15\times14 = \frac{20!}{13!}$ because it's the numbers from 20 to 1 multiplied together, divided by the numbers from 13 to 1 multiplied together.

Believe it or not, there's an even shorter form: $\frac{20!}{7!13!}$ is written as ${}^{20}C_7$ or $\binom{20}{7}$

$${}^nC_r = \binom{n}{r} = \frac{n!}{r!(n-r)!}$$

So, to finish the example, $(1+x)^{20} = 1+\frac{20}{1}x+\frac{20\times19}{1\times2}x^2+\frac{20\times19\times18}{1\times2\times3}x^3+.... = 1+20x+190x^2+1140x^3+....$

Binomial Expansions

Example: What is the term in x^5 in the expansion of $(1 - 3x)^{12}$?

The term in x^5 will be as follows:

$$\frac{12\times11\times10\times9\times8}{1\times2\times3\times4\times5}(-3x)^5$$

$$=\frac{12!}{5!7!}(-3)^5x^5 = -\frac{12!}{5!7!}\times3^5x^5 = -192456x^5$$

Watch out — the -3 is included here with the x.

Note that $(-3)^{even}$ will always be positive and $(-3)^{odd}$ will always be negative.
Here's another tip — the digits on the bottom of the fraction should always add up to the number on the top.

Some Binomials contain More Complicated Expressions

The binomials on the last page all had a 1 in the brackets — things get tricky when there's a number other than 1.
Don't panic, though. The method is the same as before once you've done a bit of factorising.

Example: What is the coefficient of x^4 in the expansion of $(2 + 5x)^7$?

Factorising $(2+5x)$ gives $2(1+\frac{5}{2}x)$

So, $(2+5x)^7$ gives $2^7(1+\frac{5}{2}x)^7$

It's really easy to forget the first bit (here it's 2^7) — you've been warned...

$$(2+5x)^7 = 2^7(1+\frac{5}{2}x)^7$$

$$=2^7[1+7\left(\frac{5}{2}x\right)+\frac{7\times6}{1\times2}\left(\frac{5}{2}x\right)^2+\frac{7\times6\times5}{1\times2\times3}\left(\frac{5}{2}x\right)^3+\frac{7\times6\times5\times4}{1\times2\times3\times4}\left(\frac{5}{2}x\right)^4+...]$$

Here's the one you want.

The coefficient of x^4 will be $2^7\times\frac{7!}{4!3!}(\frac{5}{2})^4=175000$

Don't forget the 2^7.

So, there's no need to work out all of the terms.
In fact, you could have gone directly to the term in x^4 by using the method on page 24.

Note: The question asked for the coefficient of x^4 in the expansion, so don't include any x's in your answer.
If you'd been asked for the term in x^4 in the expansion, then you should have included the x^4 in your answer.
Always read the question very carefully.

1) Sample exam question:

a) Write down the first four terms in the expansion of $(1 + ax)^{10}$, $a > 0$. [2 marks]

b) Find the coefficient of x^2 in the expansion of $(2 + 3x)^5$. [2 marks]

c) If the coefficients of x^2 in both expansions are equal, find the value of a. [2 marks]

Pascal was fine at maths but rubbish at music — he only played the triangle...

You can use your calculator to work out these tricky fractions — you use the ^nC_r button (though it could be called something else on your calculator). So to work out $^{20}C_7$, press '20', then press the ^nC_r button, then press '7', and then finish with '='. Now work out $^{15}C_7$ and $^{15}C_8$ — you should get the same answers, since they're both $\frac{15!}{7!8!}$.

Laws of Indices

You use the laws of indices a helluva lot in maths — when you're integrating, differentiating and ...er.. well loads of other places. So take the time to get them sorted now.

The Laws of Indices are the same thing as The Power Laws

Three mega-important Laws of Indices

You won't get far without knowing these three rules.

$$a^m \times a^n = a^{m+n}$$

If you multiply two numbers — you add their powers.

$a^2 a^3 = a^5$

$x^{-2} x^5 = x^3$

$p^{\frac{1}{2}} \cdot p^{\frac{1}{4}} = p^{\frac{3}{4}}$

$(a+b)^2 (a+b)^5 = (a+b)^7$

$y \cdot y^3 = y^4$

$ab^3 \cdot a^2 b = a^3 b^4$

The dot just means 'multiplied by'.

Since $y = y^1$.

Add the powers of a and b separately.

$$\frac{a^m}{a^n} = a^{m-n}$$

If you divide two numbers — you subtract their powers.

$\frac{x^5}{x^2} = x^3$

$\frac{x^{\frac{3}{4}}}{x} = x^{-\frac{1}{4}}$

$\frac{x^3 y^2}{x y^3} = x^2 y^{-1}$

Subtract the powers of x and y separately.

$$(a^m)^n = a^{mn}$$

If you have a power to the power of something else — multiply the powers together.

$(x^2)^3 = x^6$

$\{(a+b)^3\}^4 = (a+b)^{12}$

$(ab^2)^4 = a^4 (b^2)^4 = a^4 b^8$

This power applies to both bits inside the brackets.

Other important stuff about Indices

Learn to enjoy this stuff — you'll get your money's worth out of it, that's for sure.

$$a^{\frac{1}{m}} = \sqrt[m]{a}$$

You can write roots as powers...

$x^{\frac{1}{5}} = \sqrt[5]{x}$

$4^{\frac{1}{2}} = \sqrt{4} = 2$

$125^{\frac{1}{3}} = \sqrt[3]{125} = 5$

$$a^{\frac{m}{n}} = \sqrt[n]{a^m} = \left(\sqrt[n]{a}\right)^m$$

A power that's a fraction like this is the root of a power — or the power of a root.

$9^{\frac{3}{2}} = \left(9^{\frac{1}{2}}\right)^3 = \left(\sqrt{9}\right)^3 = 3^3 = 27$

$16^{\frac{3}{4}} = \left(16^{\frac{1}{4}}\right)^3 = \left(\sqrt[4]{16}\right)^3 = 2^3 = 8$

It's often easier to work out the root first, then raise it to the power.

$$a^{-m} = \frac{1}{a^m}$$

A negative power means it's on the bottom line of a fraction.

$x^{-2} = \frac{1}{x^2}$

$2^{-3} = \frac{1}{2^3} = \frac{1}{8}$

$(x+1)^{-1} = \frac{1}{x+1}$

$$a^0 = 1$$

This works for any number or letter.

$x^0 = 1$

$2^0 = 1$

$(a+b)^0 = 1$

Surds

A surd is a number like $\sqrt{2}$, $\sqrt[3]{12}$ or $5\sqrt{3}$ — one that's written with the $\sqrt{\ }$ sign. They're important because you can give exact answers where you'd otherwise have to round to a certain number of decimal places.

Surds are sometimes the only way to give an *Exact Answer*

Put $x = \sqrt{2}$ into a calculator and you'll get something like
$x = 1.414213562...$ But square 1.414213562 and you get 1.999999999.

And no matter how many decimal places you use, you'll never get exactly 2. Which is why it's best to use surds to give your answers in maths (unless the question asks for a certain number of decimal places, obviously).

Rules of Surds

$$\sqrt{ab} = \sqrt{a}\sqrt{b}$$

$$\sqrt{\frac{a}{b}} = \frac{\sqrt{a}}{\sqrt{b}}$$

$$a = \left(\sqrt{a}\right)^2 = \sqrt{a}\sqrt{a}$$

There are basically *Three Rules* for using *Surds*

Examples: (i) Simplify $\sqrt{12}$ and $\sqrt{\frac{3}{16}}$. (ii) Show that $\frac{9}{\sqrt{3}} = 3\sqrt{3}$. (iii) Find $\left(2\sqrt{5}+3\sqrt{6}\right)^2$.

(i) Simplifying surds means making the number in the $\sqrt{\ }$ sign smaller, or getting rid of a fraction in the $\sqrt{\ }$ sign. ⟹ $\sqrt{12} = \sqrt{4\times3} = \sqrt{4}\times\sqrt{3} = 2\sqrt{3}$ $\qquad \sqrt{\frac{3}{16}} = \frac{\sqrt{3}}{\sqrt{16}} = \frac{\sqrt{3}}{4}$

(ii) For questions like these, you have to write a number (here, it's 3) as $3 = \left(\sqrt{3}\right)^2 = \sqrt{3}\times\sqrt{3}$. ⟹ $\frac{9}{\sqrt{3}} = \frac{3\times3}{\sqrt{3}} = \frac{3\times\sqrt{3}\times\sqrt{3}}{\sqrt{3}} = 3\sqrt{3}$

(iii) Multiply surds very carefully — it's easy to do something a bit daft... ⟹

$$\left(2\sqrt{5}+3\sqrt{6}\right)^2 = \left(2\sqrt{5}\right)^2 + 2\times\left(2\sqrt{5}\right)\times\left(3\sqrt{6}\right) + \left(3\sqrt{6}\right)^2$$
$$= \left(2^2\times\sqrt{5}^2\right) + \left(2\times2\times3\times\sqrt{5}\times\sqrt{6}\right) + \left(3^2\times\sqrt{6}^2\right)$$
$$= 20 + 12\sqrt{30} + 54 = 74 + 12\sqrt{30}$$

Remove surds from the bottom of fractions by *Rationalising the Denominator*

Rationalising a denominator — sounds scary but it just means getting rid of the surds from the bottom of a fraction.

Example: Rationalise the denominator of $\frac{1}{1+\sqrt{2}}$

Just multiply the top and bottom by the denominator (but with the opposite sign in front of the surd).

This works because: $(a+b)(a-b) = a^2 - b^2$. So the surds cancel out.

$$\frac{1}{1+\sqrt{2}}\times\frac{1-\sqrt{2}}{1-\sqrt{2}} = \frac{1-\sqrt{2}}{1^2+\sqrt{2}-\sqrt{2}-\sqrt{2}^2} = \frac{1-\sqrt{2}}{1-2} = -1+\sqrt{2}$$

Practice Questions

1) ***Simplify the following: a)*** $m^3 \times m^7$ ***b)*** $r^5 \div r^3$ ***c)*** $(s^4)^7$

2) ***Simplify: a)*** $\sqrt{54}$ ***b)*** $\sqrt{108}$ ***c)*** $\sqrt{\frac{5}{81}}$

3) ***Show that*** $\frac{1}{2}\sqrt{2} = \sqrt{2}$

4) ***Rationalise the denominator of: a)*** $\frac{3}{1-\sqrt{2}}$ ***b)*** $\frac{\sqrt{5}+1}{\sqrt{5}-1}$

Make sure you get all this stuff in surd your head......

What can I say about indices apart from easy, easy, easy... (But just make sure you can do them without even needing to think about them.) Surds, well, they're a bit more fiddly — and you'll need to get plenty of practice. It's all too easy to do them in a rush and end up writing nonsense like $\sqrt{5}\times\sqrt{5} = 25$. So stay sharp and be cool. Then the whole of surd-world will be rosy.

Logs

Skip 'change of base' (at the bottom of this page) if you're doing Edexcel, AQA B, OCR A or OCR B.

Don't be put off by your parents or grandparents telling you that logs are hard. Logarithm is just a fancy word for power, and once you know how to use them you can solve all sorts of equations.

You need to be able to **Switch** between **Different Notations**

$\log_a b = c$ means the same as $a^c = b$

That means that $\log_a a = 1$ and $\log_a 1 = 0$

The logarithm of 100 to the base 10 is 2, because 10 raised to the power of 2 is 100.

Example: Index notation: $10^2 = 100$ log notation: $\log_{10} 100 = 2$

The base goes here but it's usually left out if it's 10.

Logs can be to any base. Base 10 is the most common — and the log button on your calculator gives logs to base 10. Natural logarithms (logs to the base e) are very important in calculus — there's more about them on page 30.

Example:

Write down the values of the following:

a) $\log_2 8$ b) $\log_9 3$ c) $\log_5 5$

a) 8 is 2 raised to the power of 3
so $2^3 = 8$ and $\log_2 8 = 3$

b) 3 is the square root of 9, or $9^{½} = 3$
so $\log_9 3 = ½$

c) anything to the power of 1 is itself
so $\log_5 5 = 1$

Write the following using log notation:

a) $5^3 = 125$ b) $3^0 = 1$

You just need to make sure you get things in the right place.

a) 3 is the power or logarithm that 5 (the base) is raised to to get 125
so $\log_5 125 = 3$

b) you'll need to remember this one:
$\log_3 1 = 0$

The **Laws of Logarithms** are **Unbelievably Useful**

Whenever you have to deal with logs, you'll end up using these laws. That means it's no bad idea to learn them off by heart right now.

Laws of Logarithms

$$\log_a x + \log_a y = \log_a (xy)$$

$$\log_a x - \log_a y = \log_a \left(\frac{x}{y}\right)$$

$$\log_a x^k = k \log_a x$$

If you're doing AQA A, you also need to know how to change the base of a log:

Change of Base (AQA A only)

$$\log_a x = \frac{\log_b x}{\log_b a}$$

Example: Calculate $\log_7 4$ to 4 decimal places.

$$\log_7 4 = \frac{\log_{10} 4}{\log_{10} 7} = 0.7124. \text{ (To check: } 7^{0.7124} = 4)$$

Logs

Use the **Laws** to **Manipulate Logs**

Whenever you get some logs to mess around with, go straight for your log laws. Obvious when you think about it.

Example: Write each expression in the form $\log_a n$, where n is a number.

a) $\log_a 5 + \log_a 4$ b) $\log_a 12 - \log_a 4$ c) $2 \log_a 6 - \log_a 9$

a) Use the law of logarithms

$\log_a x + \log_a y = \log_a (xy)$

You just have to multiply the numbers together:

$\log_a 5 + \log_a 4 = \log_a (5 \times 4)$

$= \log_a 20$

b) Use the law of logarithms

$\log_a x - \log_a y = \log_a (\frac{x}{y})$

Divide the numbers:

$\log_a 12 - \log_a 4 = \log_a (12 \div 4)$

$= \log_a 3$

c) Convert $2 \log_a 6$ using the law

$\log_a x^k = k \log_a x$

$2 \log_a 6 = \log_a 6^2 = \log_a 36$

$\log_a 36 - \log_a 9 = \log_a (36 \div 9)$

$= \log_a 4$

Practice Questions

1) Write down the values of the following

(a) $\log_3 27$ *(b)* $\log_3 (^1/_{27})$ *(c)* $\log_3 18 - \log_3 2$

2) Simplify the following

(a) $\log 3 + 2 \log 5$ *(b)* $\frac{1}{2} \log 36 - \log 3$

3) Simplify $\log_b (\chi^2 - 1) - \log_b (\chi - 1)$

4) Sample exam question:

a) Write down the value of $\log_3 3$ [1 mark]

b) Given that $\log_a \chi = \log_a 4 + 3 \log_a 2$ show that $\chi = 32$ [2 marks]

It's sometimes hard to see the wood for the trees — especially with logs...

Tricky, tricky, tricky... I think of $\log_a b$ as 'the power I have to raise a to if I want to end up with b' — that's all it is. And the log laws make a bit more sense if you think of 'log' as meaning 'power'. For example, you know that $2^a \times 2^b = 2^{a+b}$ — this just says that if you multiply two numbers, you add the powers. Well, the first law of logs is saying the same thing. Any road up, even if you don't really understand why they work, make sure you know the log laws like you know your own navel.

e^x and ln x

There's usually a question included in the Exam on natural logs and exponentials.
And you'll be ready to deal with it too, if you learn the stuff on these pages.

ln x is the Inverse of e^x

e^x is an exponential with base e.

OK, maybe e^x sounds pretty nasty — but remember that e is just a number, like π is a number.

e is about 2.718

Graph of e^x

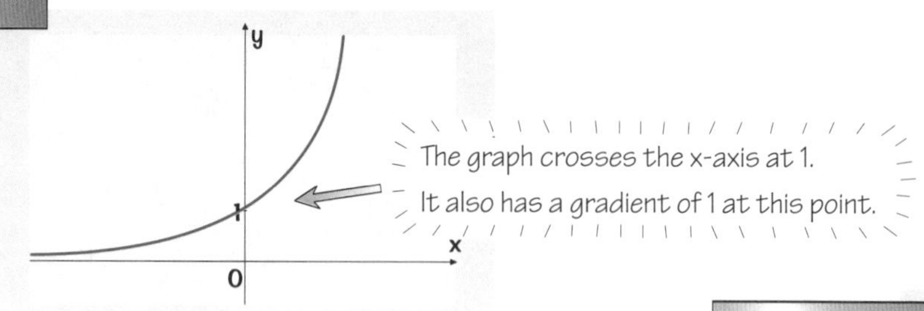

Graph of ln x

As e^x is an exponential with base e, so ln x is the inverse of e^x.

Don't panic — ln x is just the same as the logs on pages 28 and 29, but with base e.

ln x is the inverse of e^x, so if you draw it onto the same graph as e^x you can see that they mirror each other through the line y=x.
(See p6 for more info.)

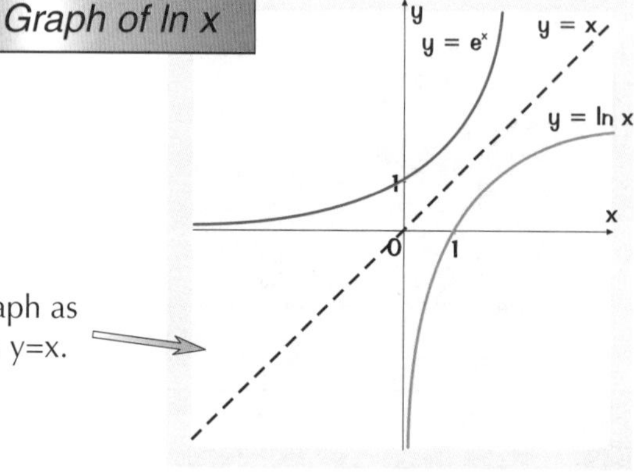

Values for e^x and ln x are Limited

Take another look at the graph of e^x above — you can see that e^x has no y-values of zero or less.

The graph of ln x is just e^x mirrored in the line y = x, so ln x can't be defined for values of x that are less than or equal to zero.

Here's a couple of special cases to learn:

As $e^1 = e$, $\ln e = 1$

As $e^0 = 1$, $\ln 1 = 0$

You can use e^x and logs to solve equations

Examples:

1) Solve $e^{4x} = 3$ to 3 decimal places.

You want x on its own, so take the natural log of both sides. $\ln e^{4x} = \ln 3$

ln x and e^x are inverse functions, so they cancel out. $4x = \ln 3$

You can now divide both sides by 4 to get x on its own. $x = \frac{\ln 3}{4}$

Remember $\frac{\ln 3}{4}$ is just a number you can find using a calculator. $x = 0.275$

I don't know about you, but I enjoyed that more than the biggest, fastest rollercoaster. You want another? OK then...

2) Solve $7 \ln x = 5$ to 3 decimal places.

You want x on its own, so begin by dividing both sides by 7. $\ln x = \frac{5}{7}$

You now need to take the exponential of both sides. $e^{\ln x} = e^{\frac{5}{7}}$

ln x and e^x are inverse functions, so they cancel out. $x = e^{\frac{5}{7}}$

Remember $e^{\frac{5}{7}}$ is just a number we can find using a calculator. $x = 2.043$

e^x and ln x

Skip the 'log graph' stuff on this page if you're doing Edexcel, AQA A, AQA B or OCR A.
Log graphs can be useful when you have to work out *a* and *b* in equations of the form $y = ax^b$ or $y = ab^x$.

Use log graphs to find Missing Numbers

Say you know that quantities *x* and *y* are linked by an equation of the form $y = ax^b$, but you don't know what *a* and *b* are. To find them, you could collect some readings and plot a graph. The reasoning goes like this...

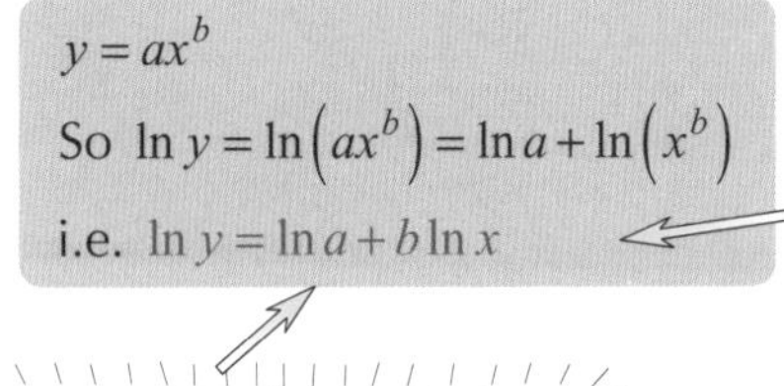

$y = ax^b$

So $\ln y = \ln(ax^b) = \ln a + \ln(x^b)$

i.e. $\ln y = \ln a + b \ln x$

This means that the graph of ln *y* against ln *x* should be a straight line — with gradient *b*, and intercept on the vertical axis equal to ln *a*.

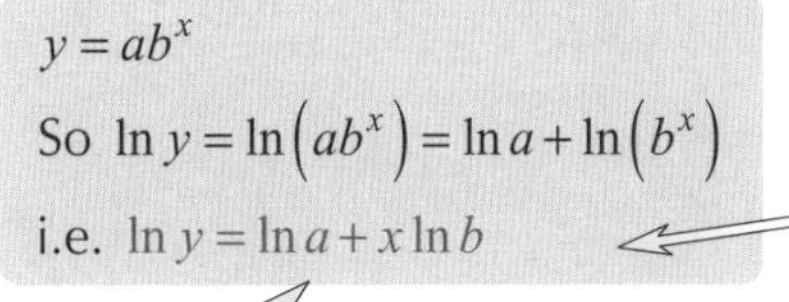

The gradient of the line is $\frac{5.38}{1.79} = 3$, so b = 3.

The intercept on the vertical axis is 0.69, so ln a = 0.69.

Therefore $a = e^{0.69} = 2$.

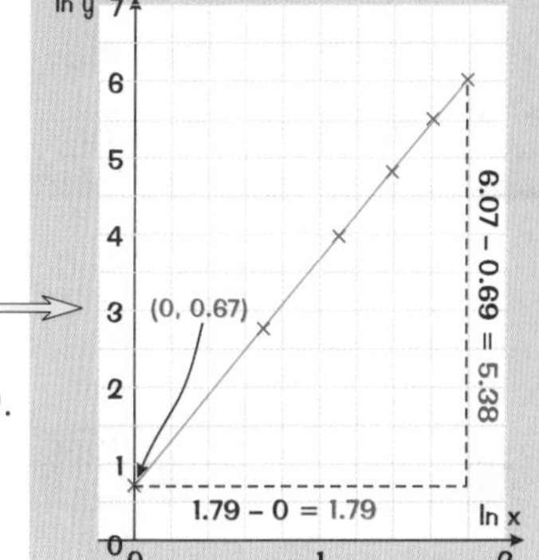

However, if *x* and *y* are linked by an equation of the form $y = ab^x$, you plot ln *y* against *x*.

$y = ab^x$

So $\ln y = \ln(ab^x) = \ln a + \ln(b^x)$

i.e. $\ln y = \ln a + x \ln b$

When you draw the graph of ln *y* against *x*, it should be a straight line with gradient ln *b* and intercept on the vertical axis equal to ln *a*.

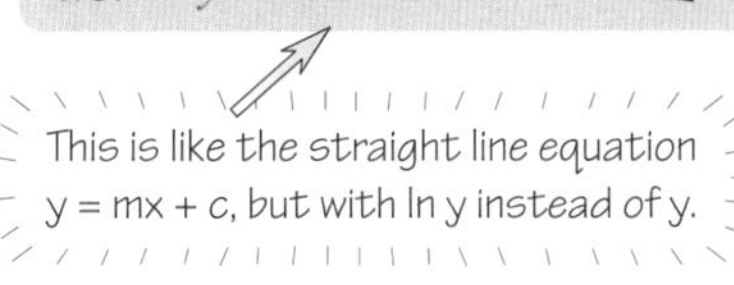

The gradient of the line is $\frac{3.47}{5} = 0.694$, so ln b = 0.694,

meaning $b = e^{0.694} = 2$.

The intercept on the vertical axis is 1.4, so ln a = 1.4.

Therefore $a = e^{1.4} = 4.06$.

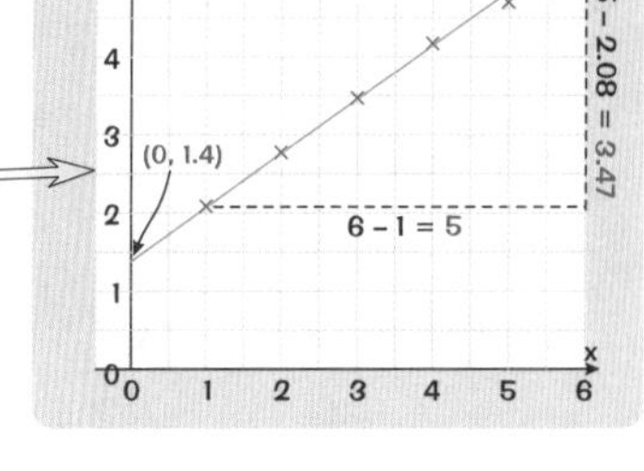

Practice Questions

1) ***Solve ln 7x = 3 to 3 decimal places.***

2) ***Calculate x to 3 decimal places where $e^{4x} = 5$.***

3) ***Solve ln 4x + ln 3 = 6 to 3 decimal places.***

4) ***Calculate x to 3 decimal places where ln 5x + ln x = 7.***

5) ***Solve x to 3 decimal places where $e^x \times e^{2x} = 6$.***

6) ***Find a and b if $y = ab^x$, and you have the following readings for x and y:***

x	1	2	3	4	5	6
y	28	196	1372	9604	67228	470596

This isn't what nutritionists mean when they start talking about e-numbers...

What's this 'e'-nonsense all about then... Well, you can take logs to any base, which sounds pretty complicated — but the base is just the little number you write after the word log. Now since the log laws work the same with any base, it often doesn't matter what base you use. Natural logs use 'e' as the base (where 'e' is about 2.718), and when you use natural logs, you write 'ln' instead of '$\log_e$'. Put like that, it doesn't seem so bad. Even I understood this eventually.

More Exponentials

Skip these two pages if you're doing AQA A or AQA B.

Okay, you've done the theory of logs and exponentials and all the stuff and all the business. Now get your calculator out.

Graphs of a^x Never Reach Zero

All the graphs of $y = a^x$ (where $a > 1$) have the same basic shape.
The graphs for $a = 2$, $a = 3$ and $a = 4$ are shown on the right.

- All the a's are greater than 1 — so y increases as x increases.
- The bigger a is, the quicker the graphs increase. The rate at which they increase gets bigger too.
- As x decreases, y decreases at a smaller and smaller rate — y will approach zero, but never actually get there.

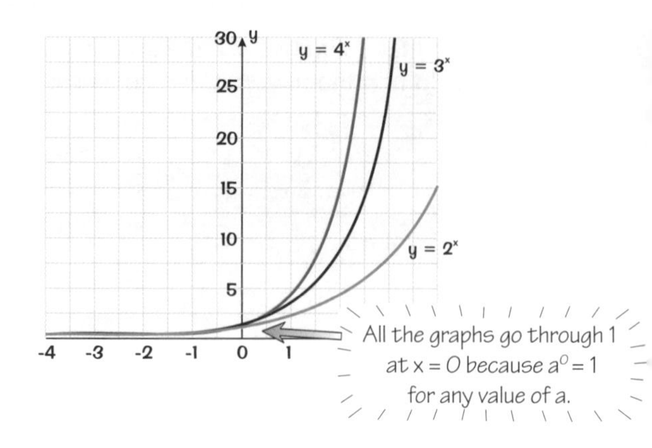

Estimate Roots Using Graphs

Example: Find x to 3 significant figures, where $3^x = 25$.

Your first step is to draw a graph to estimate where the root is. Just use $y = 3^x - 25$, and see where it crosses the axis.

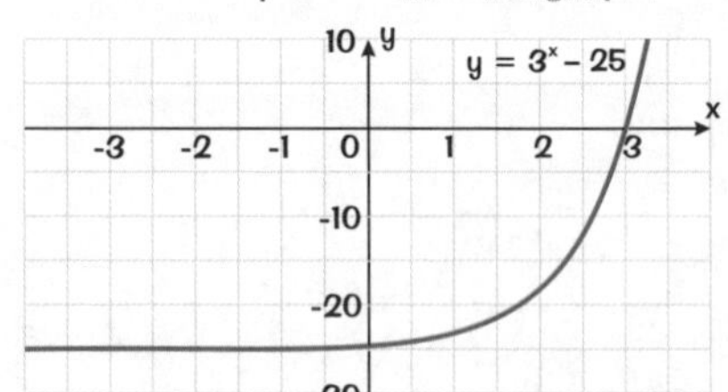

It's obviously between x = 2 and x = 3, nearer to x = 3.

$3^{2.9} = 24.2$
$3^{3.0} = 27$

Trying the next decimal point gives x between 2.9 and 3.0

$3^{2.92} = 24.7$
$3^{2.93} = 25.001$

Another decimal point on gives x between 2.92 and 2.93

It looks like the answer is x = 2.93 to 3 s.f. — but to make sure, work out: $3^{2.925} = 24.864$

This tells you the answer is definitely between 2.925 (too small) and 2.93 (too big). So **x = 2.93** to 3 s.f.

Use the Calculator Log Button Whenever You Can

Example: Use logarithms to solve the following for x, giving the answers to 4 s.f.
a) $10^x = 170$ b) $10^{3x} = 4000$ c) $7^x = 55$ d) $\log_{10}x = 2.6$ e) $2^{4x} = 80$

You've got the magic buttons on your calculator, but you'd better follow the instructions and show that you know how to use the log rules covered earlier.

a) $10^x = 170$. Taking logs to base 10 of both sides gives $x = \log_{10}170 = 2.230$.

b) $10^{3x} = 4000$. Taking logs of both sides gives $3x = \log_{10} 4000 = 3.602$, so $x = 1.201$.

c) $7^x = 55$. Take logs of both sides, and use the log rules. It doesn't matter what base you use, so why not use base 10? (Make sure you use the same base for each side though.)

$$x \log_{10}7 = \log_{10}55 \quad \text{so} \quad x = \frac{\log_{10}55}{\log_{10}7} = 2.059$$

d) $\log_{10}x = 2.6$. You've got to be able to go back the other way, so $x = 10^{2.6} = 398.1$

e) $2^{4x} = 80$. Take logs — this time why not use base e: $4x (\ln 2) = \ln 80$, so $x = \dfrac{\ln 80}{4 \ln 2} = 1.580$

More Exponentials

Exponential Growth and Decay Applies to Real-life Problems

Things like savings, radioactive decay and the growth of bacteria are real-life examples of where logs can be used to solve problems. You may well only have to solve the problems in OCR B, but you need to understand the principles in other syllabuses.

Example: The radioactivity of a substance decays by 20 per cent over a year. The initial level of radioactivity is 400. Find the time taken for the radioactivity to fall to 200 (the half-life).

$R = 400 \times 0.8^T$ where R is the level of radioactivity at time T years.
We need R = 200 so $200 = 400 \times 0.8^T$

The 0.8 comes from 1 − 20% decay.

$$0.8^T = \frac{200}{400} = 0.5.$$

$$T \ln 0.8 = \ln 0.5.$$

I've used ln, but it could be logs to any base.

$$T = \frac{\ln 0.5}{\ln 0.8} = 3.106 \text{ years}$$

Practice Questions

1) (a) Copy and complete the table for the function $y = 4^x$

x	-3	-2	-1	0	1	2	3
y							

(b) Using suitable scales, plot a graph of $y = 4^x$ for $-3 < x < 3$

(c) Use the graph to solve the equation $4^x = 20$

2) Solve the following:

(a) $10^x = 240$ (b) $\log_{10} x = 2.6$ (c) $10^{2x+1} = 1500$ (d) $4^{(x-1)} = 200$

3) Find the smallest integer P such that $1.5^P > 1\,000\,000$

4) Sample exam question:

a) Solve the following for x: $2(5^{2x}) + 2(5^x) = 12$ [3 marks]

b) The table below gives some results for an equation of the form $y = ab^x$ where a and b are both integers.

x	0	1	2	3	4
y	4	12	36	108	324

Using logarithms, find a and b. [3 marks]

If in doubt, take the natural log of something — that usually works...

The thing about exponential growth is that it's really useful, as it happens in real life all over the place. Money in a bank account earns interest at a certain percentage per year, and so the balance rises exponentially (if you don't spend or save anything). Likewise, if you got 20% cleverer for every week you studied, you'd eventually end up infinitely clever. Awesome.

Differentiating x^n

Differentiation is one of those topics that you've just got to get your head round — but luckily it's not too bad really. So relax, take a deep breath, clear your mind of any distracting thoughts, and then engrave this formula upon your heart.

The Differentiation Formula works for Any n

Differentiation

If $y = x^n$ then $\frac{dy}{dx} = nx^{n-1}$

This formula works for any n, even if it's negative or a fraction.

Example: Differentiate each function with respect to x:
a) x^{-2} b) $x^{1/2}$ c) $2x - 3x^{-1}$

a) $y = x^{-2}$. Using the rule you get:

$$\frac{dy}{dx} = nx^{n-1} = -2x^{(-2-1)} = -2x^{-3}$$

b) $y = x^{1/2}$

$$\text{so } \frac{dy}{dx} = nx^{n-1} = \tfrac{1}{2}x^{-1/2} = \frac{1}{2x^{\frac{1}{2}}} = \frac{1}{2\sqrt{x}}$$

c) $y = 2x - 3x^{-1}$

$$\frac{dy}{dx} = nx^{n-1} = 2 - (3)(-1)x^{(-1-1)} = 2 + 3x^{-2}$$

Take the terms one at a time and be careful about the minus signs.

Example: Find the gradient of the curve $y = 4x^{3/4} + x^{-1}$ at the point (1,5).

To find the gradient, first differentiate:

$$\frac{dy}{dx} = (4 \times \tfrac{3}{4})x^{-1/4} + (-1)x^{-2} = 3x^{-1/4} - x^{-2}$$

$n - 1 = \tfrac{3}{4} - 1 = -\tfrac{1}{4}$

Now substitute in x = 1:

$$\frac{dy}{dx} = (3)(1^{-1/4}) - (1^{-2}) = 3 - 1 = 2$$

You only need the x-value of the point (1,5).

Differentiating x^n

Index Notation makes Fractional Powers easier to work with

To differentiate some functions you'll need to rewrite them in index notation first.

Remember the rules of indices:

Rules of Indices

$\frac{1}{x^n} = x^{-n}$	$\sqrt[b]{(x^a)} = x^{\frac{a}{b}}$

Example:

Differentiate each function with respect to x:

a) $\sqrt[3]{x}$ b) $\frac{2}{x^3}$ c) $\frac{4}{x^2} - \frac{3}{\sqrt{x}}$

a) First rewrite in index notation: $\sqrt[3]{x} = x^{\frac{1}{3}}$

Now differentiate using the rule: $\frac{dy}{dx} = \frac{1}{3}x^{(\frac{1}{3}-1)} = \frac{1}{3}x^{-\frac{2}{3}} = \frac{1}{3\sqrt[3]{x^2}}$

b) As before, rewrite using index notation: $\frac{2}{x^3} = 2x^{-3}$

Turn the index notation back into a fraction at the end.

Differentiating gives: $\frac{dy}{dx} = (2)(-3)x^{(-3-1)} = -6x^{-4} = \frac{-6}{x^4}$

$\sqrt{x} = x^{\frac{1}{2}}$ so $\frac{1}{\sqrt{x}} = x^{-\frac{1}{2}}$

c) Each term must be put into the correct form: $\frac{4}{x^2} - \frac{3}{\sqrt{x}} = 4x^{-2} - 3x^{-\frac{1}{2}}$

Differentiate each term one by one: $\frac{dy}{dx} = (4)(-2)x^{(-2-1)} - (3)(-½)x^{(-½-1)} = -8x^{-3} + \frac{3}{2}x^{-\frac{3}{2}} = -\frac{8}{x^3} + \frac{3}{2\sqrt{x^3}}$

Practice Questions

1) Differentiate each of the following functions:

a) $y = x^{-4}$ *b) $y = 3x^{½}$* *c) $f(x) = 4x^2 - 3x^{-3}$*

2) For each curve find (i) $\frac{dy}{dx}$ and (ii) the gradient at the given point:

a) $y = 3\sqrt{x} + \frac{4}{x}$ at point (4,7) *b) $y = \frac{2}{x^2} + \frac{3}{x^3}$ at point (1,5)*

***3)** a) Verify that the graphs $y = 4 - x$ and $y = \frac{2}{\sqrt{x}} + 1$ intersect at (1,3)*

b) Differentiate $y = \frac{2}{\sqrt{x}} + 1$

c) Find the gradient of $y = \frac{2}{\sqrt{x}} + 1$ at (1,3)

I apologise now for the rant that follows...

Practice is the key. So differentiate any x^n formulas you see — make it your mission in life to differentiate. Differentiate x^n things until you can do it in your sleep. Even if you're not really sure what differentiation actually means, you'll still get marks in the Exam for knowing how to do it. And there's only one formula. So there's absolutely no excuse for not knowing it.

Differentiation of e^x and ln x

This kind of question comes up in the Exam every year. They're easy to spot — and even better, they're easy to do. So if you learn the bits and bobs on these pages, you can guarantee yourself some juicy marks.

Differentiating e^x gives e^x

Sounds weird, but it's true — e^x is the only function that doesn't change when you differentiate it.

Here are some examples:

$$y = e^x \qquad \frac{dy}{dx} = e^x$$

$$y = 2e^x \qquad \frac{dy}{dx} = 2e^x$$

$$y = -0.3e^x \qquad \frac{dy}{dx} = -0.3e^x$$

If all of AS maths were like this, it'd be a doddle.

Example: If y = ke^x, find the gradient when x = 2.

First differentiate to get the gradient: $\frac{dy}{dx} = ke^x$ ← e^x is unchanged — k just stands for an unknown number.

Now put in x = 2: $\frac{dy}{dx} = ke^2$ So the gradient is ke^2.

Example: Find the x-coordinate of the stationary point on $y = 2e^x - 2x$.

You need the gradient again: $\frac{dy}{dx} = 2e^x - 2$ Stationary points occur when $\frac{dy}{dx} = 0$ →

$$2e^x - 2 = 0$$
$$e^x = 1$$
$$x = 0$$

Skip this subsection if you're doing Edexcel P2.

But Differentiating e^{kx} gives ke^{kx}

You'll probably not be so lucky as to just get e^x — but don't worry. It's not much harder to cope with e^{kx}.

Here are a few examples:

$$y = e^{kx} \qquad \frac{dy}{dx} = ke^{kx}$$

$$y = e^{5x} \qquad \frac{dy}{dx} = 5e^{5x}$$

$$y = e^{-2x} \qquad \frac{dy}{dx} = -2e^{-2x}$$

$$y = e^{-x} \qquad \frac{dy}{dx} = -e^{-x}$$

If there's a number in front of the e^x, then that number just stays the same:

$$y = 2e^{3x}$$ ← e^{3x} differentiated as normal.

$$\frac{dy}{dx} = 2 \times 3e^{3x}$$ ← 2 stays the same.

Example: If y = $e^{2x} - 6x$, show that the x-coordinate of the stationary point is $\frac{1}{2}$ ln 3.

Start by differentiating: $\frac{dy}{dx} = 2e^{2x} - 6$

For a stationary point $\frac{dy}{dx} = 0$, so: $2e^{2x} - 6 = 0$

Rearrange to give: $e^{2x} = 3$

Take logs of both sides:

$$\ln e^{2x} = \ln 3$$
$$2x = \ln 3$$
$$x = \frac{1}{2}\ln 3$$

Differentiation of e^x and ln x

Differentiating ln x gives $1/x$

This one's not so obvious — you just have to learn it.

Here's some examples:

$$y = \ln x \qquad \frac{dy}{dx} = \frac{1}{x}$$

$$y = 3\ln x \qquad \frac{dy}{dx} = 3 \times \frac{1}{x} = \frac{3}{x}$$

$$y = \frac{1}{4}\ln x - 7x \qquad \frac{dy}{dx} = \frac{1}{4} \times \frac{1}{x} - 7 = \frac{1}{4x} - 7$$

And Differentiating ln kx also gives $1/x$

Because ln 4x = ln 4 + ln x, the graph of ln 4x is just like the graph of ln x but shifted up ln 4 units.

But its gradient hasn't changed, so $\frac{dy}{dx}$ stays the same.

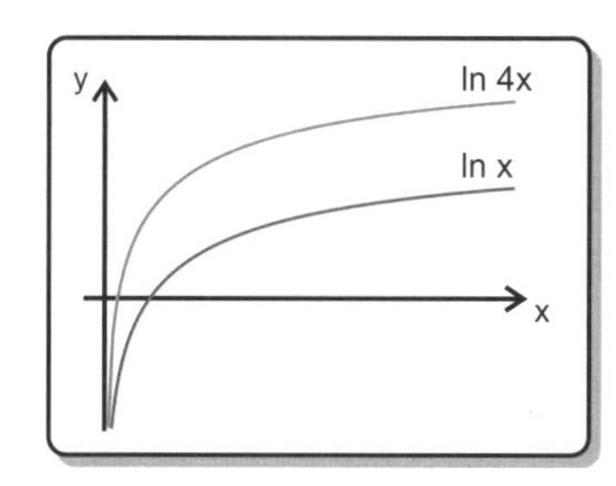

But watch out — if the constant is in front of the log, it remains unaffected.

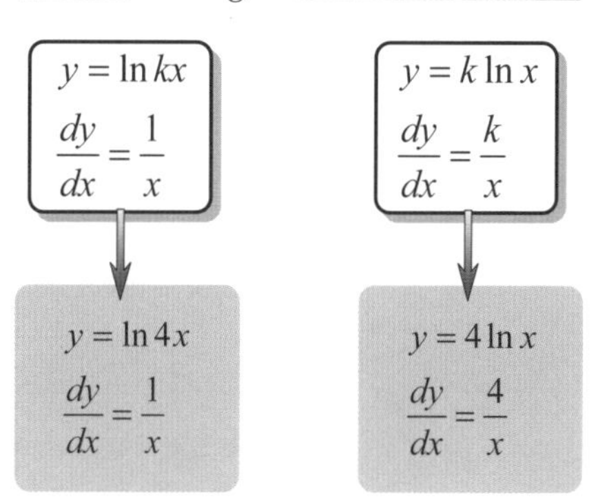

Example: Differentiate $y = 2\ln 5x$.

$$\frac{dy}{dx} = 2 \times \frac{1}{x} = \frac{2}{x}$$

2 is unchanged

ln 5x differentiates to $\frac{1}{x}$

Example: If $y = 8\ln 3x - x^2$, find the gradient of the curve when $x = 4$.

Differentiate as usual: $\frac{dy}{dx} = 8 \times \frac{1}{x} - 2x = \frac{8}{x} - 2x$

Now when $x = 4$, $\frac{dy}{dx} = \frac{8}{4} - 2 \times 4 = -6$

Practice Questions

1) If $y = e^x - \ln 4x$ and $x > 0$, find $\frac{dy}{dx}$ and show that the curve has a stationary point at $xe^x - 1 = 0$

2) Sample exam question:

If $f(x) = 4e^x + 3\ln 2x$, evaluate $f'(1)$, giving your answer in terms of e. [4 marks]

Uh oh — the maths-speak is starting to get everywhere...

If you're the curious type (and there's nothing wrong with not being curious, by the way), then the rules for differentiating e^{kx} and ln kx can be explained using the Chain Rule on pages 44 and 45. If you're not curious, just learn them off by heart.

Second Order Derivatives

Skip these two pages if you're doing Edexcel or OCR A.

You're probably an expert at differentiating to find stationary points by now. But do you lie awake at night agonising over whether they're maximums or minimums? If so, these two pages should solve the thorny problem.

Finding **Second Order Derivatives** just means **Differentiating y Twice**

You can find the gradient of the gradient by differentiating the gradient again. This is $\frac{d}{dx}(\frac{d}{dx}(y))$ or $\frac{d^2y}{dx^2}$.

Example:

$$y = 5x^3 - x^4$$
$$\frac{dy}{dx} = 15x^2 - 4x^3$$
$$\frac{d^2y}{dx^2} = 30x - 12x^2$$

$$y = \ln 4x - 3x^2$$
$$\frac{dy}{dx} = \frac{1}{x} - 6x \quad = x^{-1} - 6x$$
$$\frac{d^2y}{dx^2} = -x^{-2} - 6$$

If you start with **f(x)**, differentiating twice is **f''(x)**

The examiners are pretty likely to get you to muck around with e^x and ln x (because they're harder).

Example:

$$f(x) = 3\ln 2x$$
$$f'(x) = 3 \times \frac{1}{x}$$
$$= 3x^{-1}$$
$$f''(x) = -3x^{-2}$$

$$f(x) = 2e^{-x} + 4x$$
$$f'(x) = -2e^{-x} + 4$$
$$f''(x) = 2e^{-x}$$

f''(x) tells you whether you've got a **Maximum** or a **Minimum**

You can find the stationary point using $\frac{dy}{dx}$ — but it's $\frac{d^2y}{dx^2}$ that tells you what type it is.

Max or Min?

If $\frac{d^2y}{dx^2} > 0$ you have a minimum point

If $\frac{d^2y}{dx^2} < 0$ you have a maximum point

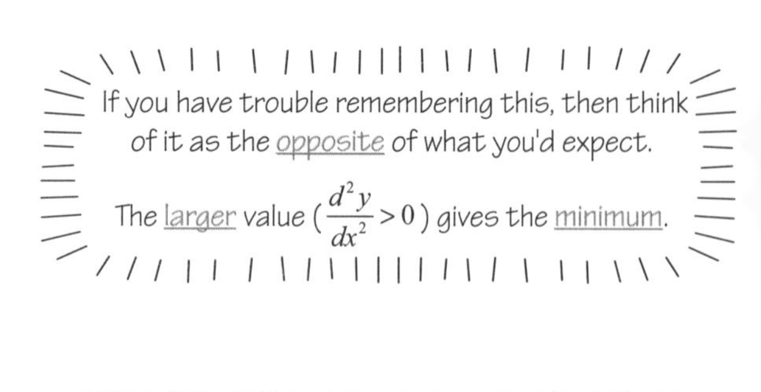

Example:

Determine the nature of the stationary point of $y = 3x^2 - 4$.

$\frac{dy}{dx} = 6x$ — stationary point is when $\frac{dy}{dx} = 0$

$0 = 6x$ so the stationary point is at $x = 0$

$\frac{d^2y}{dx^2} = 6$

$\frac{d^2y}{dx^2}$ is always greater than 0 so you get a minimum, no matter what values of x you have.

Determine the nature of the stationary point of $f(x) = 4e^{2x} - 8x$.

So $f'(x) = 8e^{2x} - 8$

There's a stationary point when $f'(x) = 0$

So $0 = 8e^{2x} - 8$

$e^{2x} = 1$ so $x = 0$ ← $e^0 = 1$

$f''(x) = 16e^{2x}$

When $x = 0$, $f''(x) = f''(0) = 16e^0 = 16$

$f''(x) > 0$ so this is a minimum too.

Second Order Derivatives

If there's More than One stationary point, Test them One by One

You might find that you have two stationary points or maybe three — just test each one.
Watch out though — if the domain is restricted, you might have fewer stationary points than you think.

Example: Determine the nature of the stationary points of $y = \ln 3x - x^2,\ x > 0$

The domain is restricted to positive values of x.

$$\frac{dy}{dx} = \frac{1}{x} - 2x = 0$$

Stationary point when $\frac{dy}{dx} = 0$

$$2x = \frac{1}{x}$$

$$x^2 = \frac{1}{2}$$

$$x = \frac{1}{\sqrt{2}}$$

$x > 0$ so $x = -\frac{1}{\sqrt{2}}$ can't be an answer.

Now differentiate again:

$$\frac{d^2y}{dx^2} = -x^{-2} - 2$$

It's easiest to write $\frac{1}{x}$ as x^{-1} before you differentiate it.

When $x = \frac{1}{\sqrt{2}}$, $\frac{d^2y}{dx^2} = -4$

This is < 0 so the stationary point's a maximum.

Note for AQA A only: You're allowed to use a graphic calculator, so you can check your answer by drawing $y = \ln 3x - x^2$ and pressing G-solv and then Max to see where the maximum is.

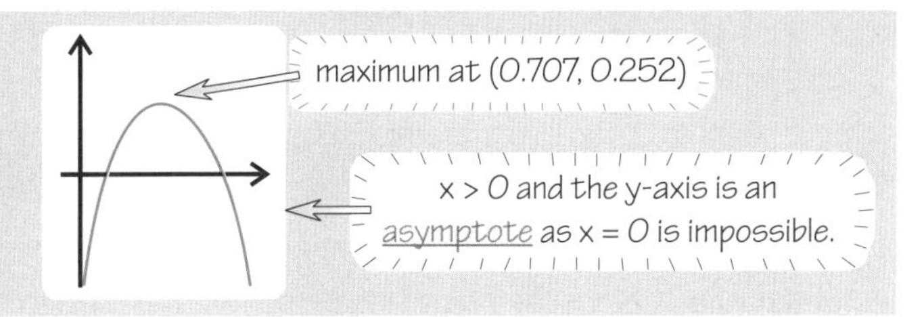

Practice Questions

1) Find the coordinates of the stationary point of the graph $y = e^{-2x} + 2x$.

2) For the equation in 1), Find $\frac{d^2y}{dx^2}$ and hence determine the nature of the stationary point.

3) Sample exam question:

A curve has the equation $y = x^2 - \ln 3x,\ x > 0$. The curve has one stationary point P.

a) Show that the x-coordinate of P is $\frac{1}{\sqrt{2}}$ [4 marks]

b) Find the value of $\frac{d^2y}{dx^2}$ at P [3 marks]

c) State, giving a reason, whether P is a maximum or a minimum point on the curve. [1 mark]

This stuff will derive you mad if you let it...

When I were a lad, I kept repeating MINUSMAX, MINUSMAX, MINUSMAX to myself. It stuck in my head pretty well, and now I want to pass this on... It means that if the second derivative is MINUS, you've got a MAX. Not a particularly clever memory aid, but it did the job for me. And just in case you were confused — yes, f''(x) does mean the same as $\frac{d^2y}{dx^2}$.

Tangents and Normals

Skip these two pages if you're doing OCR A, OCR B or AQA B.

For P1 you had to differentiate functions to find tangents and normals to a curve.
For P2 you have to do exactly the same — only the equations of the curves are a little bit trickier.

When a **Tangent Meets a Curve** they have the **Same Gradient**

You can differentiate the equation of the curve to find the gradient.
Then all you need to do is to use $\mathbf{y = mx + c}$ to find the equation of the tangent.

Example: Find the equation of the tangent to the curve $\mathbf{y = 0.5e^x - 2}$ where it cuts the y-axis.

A sketch will make it easier to understand:

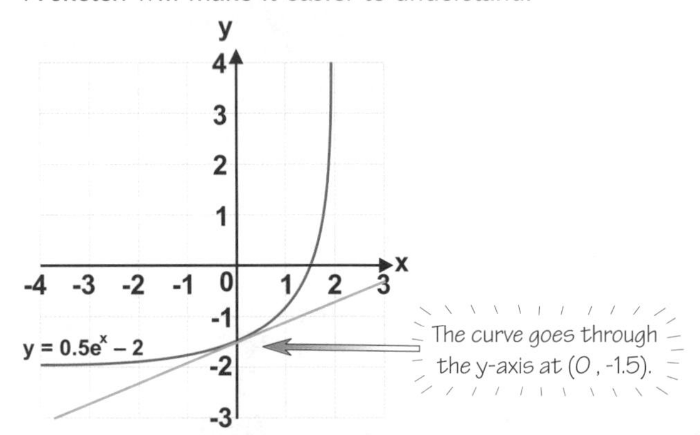

You need to differentiate to find the gradient:

$$\frac{dy}{dx} = 0.5e^x$$

x = 0 where it cuts the y-axis, so plug that in:

$$\frac{dy}{dx} = 0.5e^0$$

$e^0 = 1$

$$\frac{dy}{dx} = 0.5$$

So the tangent has gradient 0.5. You already know the intercept is at (0, -1.5) so you can write down the equation of the tangent:

$$y = 0.5x - 1.5$$

The **Normal** to a **Curve** is at **Right Angles** to the **Tangent**

The tangent to a curve at a point is the straight line with the same gradient at that point.
The normal to a curve is the line that's perpendicular (at right angles) to the tangent.

Gradient of tangent × gradient of normal = -1

Example: Find the equation of the normal to the curve $\mathbf{y = \ln 2x + 1}$ at the point where x = 2.

Draw a sketch of the graph and add the tangent and normal so that you can see the problem.

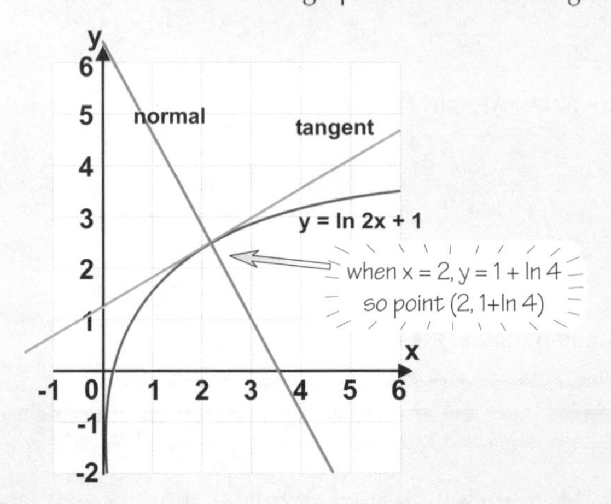

To find the gradient of the normal, first differentiate: $\frac{dy}{dx} = \frac{1}{x}$

When x = 2: $\frac{dy}{dx} = \frac{1}{2}$

Gradient of the normal is $-1 \div \frac{1}{2} = -2$

Using $\mathbf{y - y_1 = m(x - x_1)}$ you can write the equation of the normal:

$$y - (1 + \ln 4) = -2(x - 2)$$

$$y = -2x + 5 + \ln 4$$

Tangents and Normals

Watch out for Tangents and Normals at Different Points

Some questions will give you a tangent and a normal to find — but not necessarily at the same point.

Example:

The diagram shows the graph of $y = \ln x$. The normal at $x = 2$ and the tangent at $x = 1$ are also shown. Find the coordinates of their point of intersection.

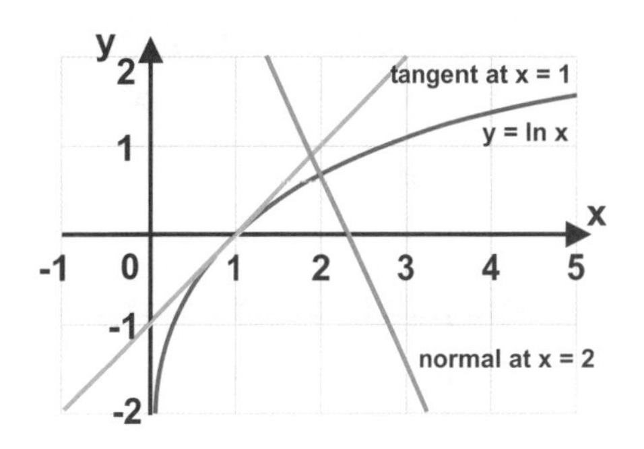

To find the point of intersection you need the equation of each straight line.

First differentiate to find the gradient: $\frac{dy}{dx} = \frac{1}{x}$

at x = 1: $\frac{dy}{dx} = 1$

Then use $y - y_1 = m(x - x_1)$: $y - 0 = 1(x - 1)$

So the equation of the tangent is: $y = x - 1$

at x = 2: $\frac{dy}{dx} = \frac{1}{2}$

Gradient of normal $= -1 \div \frac{1}{2} = -2$

Use $y - y_1 = m(x - x_1)$: $y - \ln 2 = -2(x - 2)$

So the equation of the normal is: $y = -2x + 4 + \ln 2$

To find the point of intersection you need to solve the simultaneous equations $\mathbf{y = x - 1}$ and $\mathbf{y = -2x + 4 + \ln 2}$.

Both right-hand sides are equal to y: $x - 1 = -2x + 4 + \ln 2$

Get x alone on one side: $3x = 5 + \ln 2$

$$\mathbf{x = \frac{1}{3}(5 + \ln 2)}$$

Then sub x in to get y: $y = x - 1$

$$\mathbf{so\ y = \frac{1}{3}(2 + \ln 2)}$$

So the point of intersection is: $\left(\frac{1}{3}[5 + \ln 2],\ \frac{1}{3}[2 + \ln 2]\right)$

Looks complicated — but it's just the x and y values put together.

Practice Questions

1) ***Find the equation of the tangent to the curve $y = \frac{2}{x}$ at the point (1,2).***
2) ***Find the equation of the normal to the curve $y = \sqrt{x}$ at the point (4,2).***
3) ***Sample exam question:***

The curve with equation $y = e^x - 4x$ meets the y-axis at P. Q is the point on the curve with x-coordinate ln 6. The normal to the curve at P and the tangent to the curve at Q meet at R.

a) Write down the coordinates of P and Q [2 marks]

b) Show that the normal at P has equation $3y = x + 3$ [4 marks]

c) Prove that the x-coordinate of R is $\frac{3}{5}(6 \ln 6 - 5)$ and find the y-coordinate. [6 marks]

Normal and Maths — two words I don't usually say in the same sentence...

The good news about these two pages is that there's hardly anything new on them. You learnt how to find a tangent and normal in P1. And the differentiation is the same as in the first part of this section. All I need say is make sure you can do it.

Product and Quotient Rules

Skip these two pages if you're doing Edexcel, OCR A, AQA A or AQA B.

More differentiation, and this time it's the product rule and the quotient rule. Luckily, they're a piece of cake.

Use the Product Rule to Differentiate things Multiplied together

The product rule is for differentiating things that are made up of smaller bits multiplied together.

$$\text{If } y = u(x)\times v(x)\text{, then: } \frac{dy}{dx} = \frac{du}{dx}v + u\frac{dv}{dx}$$

Or to put that in a way that's slightly easier to remember:

Remember it as "u-dash v plus u v-dash".

$$\text{If } y = uv\text{, then: } \frac{dy}{dx} = u'v + uv'$$

where the dashes mean you're differentiating that bit.

Example: Find $\frac{dy}{dx}$ if $y = (x^2-2)(x^2+2x+3)$.

You could just multiply out the brackets and differentiate what you get, but that's a pain.
It's quicker and easier to use the product rule.

1. y is two functions multiplied together — call them u and v: $u = x^2 - 2$ and $v = x^2 + 2x + 3$
2. Differentiate u and v: $\frac{du}{dx} = u' = 2x$ and $\frac{dv}{dx} = v' = 2x + 2$
3. Now use the product rule: You take the derivative of the first function and multiply it by the second function, then add the first function multiplied by the derivative of the second.

$$\text{So } \frac{dy}{dx} = 2x(x^2+2x+3) + (x^2-2)(2x+2)$$
$$= 2x^3 + 4x^2 + 6x + 2x^3 + 2x^2 - 4x - 4 = 4x^3 + 6x^2 + 2x - 4$$

Bob's your uncle.

Use the Quotient Rule to Differentiate things that are Divided

The quotient rule is pretty similar, but it's for differentiating one function divided by another.

$$\text{If } y = \frac{u(x)}{v(x)}\text{, then: } \frac{dy}{dx} = \frac{\frac{du}{dx}v - u\frac{dv}{dx}}{v^2} \quad \text{or} \quad \frac{dy}{dx} = \frac{u'v - uv'}{v^2}$$

This is "u-dash v minus u v-dash, all over v-squared".

Example: Find $\frac{dy}{dx}$ if $y = \frac{x}{e^{2x}}$.

Or you could write this as $y = xe^{-2x}$ and use the product rule — you'd get the same answer.

Take it nice and slowly...

1. Write down what u and v are: $u = x$ and $v = e^{2x}$
2. Differentiate u and v: $\frac{du}{dx} = u' = 1$ and $\frac{dv}{dx} = v' = 2e^{2x}$
3. Now use the quotient rule: Take the derivative of the top function (u') and multiply it by the bottom function (v), then subtract the top function (u) multiplied by the derivative of the bottom one (v'). Finally, divide by the square of the bottom function (v^2).

$$\text{So } \frac{dy}{dx} = \frac{1e^{2x} - x2e^{2x}}{(e^{2x})^2} = \frac{e^{2x}(1-2x)}{(e^{2x})^2} = \frac{(1-2x)}{e^{2x}}$$

Product and Quotient Rules

Use the *Product* and *Quotient Rules* for finding *Gradients*

The product and quotient rules are just ways to differentiate things — so you can use them to find gradients.

Example: Find the gradient of the curve $y = \frac{x}{\ln x}$ at the point where $x = e$.

Let $u = x$ and $v = \ln x$

Then $\frac{du}{dx} = u' = 1$ and $\frac{dv}{dx} = v' = \frac{1}{x}$, which means that $\frac{dy}{dx} = \frac{1 \ln x - x\frac{1}{x}}{(\ln x)^2} = \frac{\ln x - 1}{(\ln x)^2}$

Now stick $x = e$ in the derivative, and you find that the gradient equals $\frac{\ln e - 1}{(\ln e)^2} = \frac{1-1}{1^2} = 0$, i.e. it's a turning point.

Example: Find the turning point of the graph of $y = xe^x$. Is this a maximum or a minimum?

To find turning points, you set the derivative equal to zero — so you need to differentiate. Use the product rule.

Let $u = x$ and $v = e^x$. Then $\frac{du}{dx} = u' = 1$ and $\frac{dv}{dx} = v' = e^x$

Product Rule: $\frac{dy}{dx} = 1e^x + xe^x = (1+x)e^x$.

So $\frac{dy}{dx} = 0$ when $x = -1$ (since e^x is never 0).

Now you need to differentiate again to see if it's a maximum or a minimum. It's another product, so write $\frac{dy}{dx} = uv$.

Then $u = (1+x)$ and $v = e^x$. This means $\frac{du}{dx} = u' = 1$ and $\frac{dv}{dx} = v' = e^x$

So $\frac{d^2y}{dx^2} = 1e^x + (1+x)e^x = (2+x)e^x$.

At $x = -1$, this equals $(2+(-1))e^{-1} = \frac{1}{e} > 0$, so this is a minimum.

Practice Questions

1) ***Differentiate the following using the product rule:*** a) x^3e^{2x} b) $x \ln x$

2) ***Differentiate the following using the quotient rule:*** a) $\frac{x^2}{x+1}$ b) $\frac{e^x}{\ln x}$

3) ***Sample exam question:***

A curve has equation $y = \frac{e^x}{x^2}$ $(x \in \mathbb{R}, x \neq 0)$.

Find where the graph has its turning point(s). [5 marks]

Product and quotient rules — okay...

The product and quotient rules... Ahh, bless 'em — they're lovely. They have really scary-looking formulas, but when it comes down to it, they're not too bad at all. You just have to remember that the product rule has a plus sign in it (i.e. the two p's stick together), while the quotient rule has a minus sign, plus an extra v^2 on the bottom line.

Chain Rule

Skip these two pages if you're doing Edexcel, AQA A or AQA B.

To differentiate $y = (2x - 3)^{11}$ you *could* multiply it out and then differentiate each term in the usual way. Unfortunately, that could take a while. Luckily there's a much easier way of doing it...

Use the *Chain Rule* to differentiate *Composite Functions*

The Chain Rule

$$\frac{dy}{dx} = \frac{dy}{du} \times \frac{du}{dx}$$

How it works:

For example, $\mathbf{y = (2x - 3)^{11}}$ is a composite function. It's made up of two functions, $\mathbf{2x - 3}$ and $\mathbf{(\quad)^{11}}$.

Just replace the inner function (the function in the bracket) by u: $y = u^{11}$

Then differentiate the new function with respect to u: $\frac{dy}{du} = 11u^{10}$

And the chain rule also requires differentiating u with respect to x: $\frac{du}{dx} = 2$

Remember $u = 2x - 3$

Finish off by changing u back again.

Then apply that lovely chain rule: $\frac{dy}{dx} = \frac{dy}{du} \times \frac{du}{dx} = 11u^{10} \times 2 = 22(2x - 3)^{10}$

Example:

1) Differentiate $y = (3x - 5)^4$

Put $u = 3x - 5$

so $\frac{du}{dx} = 3$

$y = u^4$ so $\frac{dy}{du} = 4u^3$

$\frac{dy}{dx} = \frac{dy}{du} \times \frac{du}{dx} = 4u^3 \times 3 = 12(3x - 5)^3$

u is the bit in the brackets.

Change **u** back to finish off.

2) Differentiate $y = (x^2 + 2)^3$

Put $u = x^2 + 2$

so $\frac{du}{dx} = 2x$

$y = u^3$ so $\frac{dy}{du} = 3u^2$

$\frac{dy}{dx} = \frac{dy}{du} \times \frac{du}{dx} = 3u^2 \times 2x = 6x(x^2 + 2)^2$

Don't Panic if you have to use the *Chain Rule* with *Roots*

Composite functions involving roots look tricky — just remember to use index notation and it's a walk in the park.

EXAMPLE: If $y = \sqrt{4x - 2}$, find $\frac{dy}{dx}$.

Make **u** the bit inside the root sign.

If $u = 4x - 2$ then $\frac{du}{dx} = 4$

Write $u^{1/2}$ instead of $\sqrt{u}$.

$y = u^{1/2}$ so $\frac{dy}{du} = \frac{1}{2}u^{-1/2}$

Change the **u** back and write it all in terms of square roots again.

$\frac{dy}{dx} = \frac{dy}{du} \times \frac{du}{dx} = \frac{1}{2}u^{-1/2} \times 4 = 2(4x - 2)^{-1/2} = \frac{2}{\sqrt{4x - 2}}$

Chain Rule

You still have to do the *Same Old Tricks* with the *Chain Rule*

Just because you're using a clever technique, don't think you can forget about all the other stuff when you differentiate. You've still got to be able to find tangents and normals and stationary points using the chain rule.

Example: Differentiate $y = \frac{1}{x+4}$ and then find the stationary points on the curve $y = \frac{1}{x+4} + x$.

To differentiate $\frac{1}{x+4}$ you need to write it in the index form $\mathbf{y = (x + 4)^{-1}}$ and use the chain rule:

$u = x + 4$ so $\frac{du}{dx} = 1$ ← u is the part in brackets.

$y = u^{-1}$ so $\frac{dy}{du} = -u^{-2}$

$$\frac{dy}{dx} = \frac{dy}{du} \times \frac{du}{dx} = -u^{-2} \times 1 = \frac{-1}{(x+4)^2}$$

Use the answer you just got to differentiate $y = \frac{1}{x+4} + x$: $\frac{dy}{dx} = \frac{-1}{(x+4)^2} + 1$

At stationary points $\frac{dy}{dx} = 0$

$$\frac{-1}{(x+4)^2} + 1 = 0$$

$$\frac{1}{(x+4)^2} = 1$$

$(x + 4)^2 = 1$ ← Take the square root of each side — there are two possible values.

$(x + 4) = \pm 1$ so $x = -3$ and $x = -5$

i.e. there are stationary points at (-3, -2) and (-5, -6). ← Sub x into the original equation to find the y-coordinates.

Example: Find the equation of the tangent to the curve $\mathbf{y = (x^2 + 3x)^2}$ at (-2, 4).

To find the gradient of the tangent, differentiate and put in x = -2:

$y = (x^2 + 3x)^2$

$u = (x^2 + 3x)$ $\quad \frac{du}{dx} = 2x + 3$

$y = u^2$ $\quad \frac{dy}{du} = 2u$

Using the chain rule:

$\frac{dy}{dx} = 2u \times (2x + 3) = 2(2x + 3)(x^2 + 3x)$

When x = -2: $\frac{dy}{dx} = 2 \times -1 \times -2 = 4$ ← This is the gradient.

Using $y - y_1 = m(x - x_1)$:

$y - 4 = 4(x - -2)$

$y = 4x + 12$

Practice Questions

1) Differentiate each function:

a) $y = (x + 2)^7$ *b)* $y = (3x^2 + 5x)^4$ *c)* $y = \sqrt[3]{2x - 5}$ *d)* $y = \frac{1}{(4x+5)^3}$

2) Find the normal to the curve $\sqrt{(x+5)}$ ***at the point where x = -1.***

3) Sample exam question:

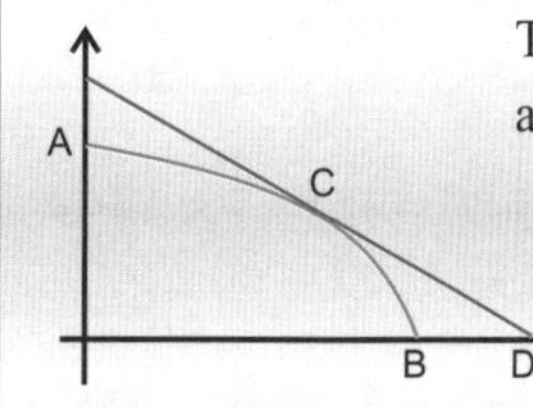

The graph shows part of $y = \sqrt{(9 - 2x)}$ for positive values of x. The curve meets the axes at A and B. The x-coordinate at the point C on the curve is 2.5. The tangent at C meets the x-axis at D.

a) Write down the coordinates of A and B. [2 marks]

b) Show that $\frac{dy}{dx} = \frac{-1}{\sqrt{(9-2x)}}$ [3 marks]

c) Find the coordinates of D. [3 marks]

Oh, my love, my darling, I've hungered for more maths — 'Un-Chain-Ruled Melody'...

It's best to take your time with the Chain Rule at first. Write down what u is, then differentiate it. Write down what y is in terms of u, then differentiate that. Then multiply the bits together to get the derivative of what you started with.

Connected Rates of Change

Skip these two pages if you're doing Edexcel, AQA A or AQA B.

Connected Rates of Change sounds a bit scary. But it just means using the Chain Rule — though you might have to use letters other than x, y and u. And when you think of it like that, it stops being quite so scary.

The chain rule doesn't have to contain y, x and u

Any letters will do — which means you can use the chain rule to solve lots of rates of change questions.

For example, instead of $\frac{dy}{dx} = \frac{dy}{du} \times \frac{du}{dx}$ you could write $\frac{dy}{dt} = \frac{dy}{dx} \times \frac{dx}{dt}$

$\frac{dy}{dt}$ is the rate of change of y with respect to t

These could be any letter as long as they're the same.

t stands for time.

Connected Rates of Change Questions are often about Area or Volume

Examiners love asking you to work out rates of change of areas or volumes with a length or radius.

A spherical balloon is inflated so that its volume increases by 5 cm³/s. Find the rate of increase of the radius when the volume is 45 cm³.

The volume of a sphere, V, in terms of the radius, r, is given by $V = \frac{4}{3}\pi r^3$.

So differentiate with respect to r: $\frac{dV}{dr} = 4\pi r^2$

You also need the rate of increase of volume: $\frac{dV}{dt} = 5 \text{ cm}^3\text{/s}$ ← This was given in the question.

When V = 45 cm³: $r = \sqrt[3]{\frac{1}{\pi} \times \frac{3}{4} \times 45} = 2.21 \text{ cm}$

Rewrite the chain rule with the letters you're using: $\frac{dV}{dt} = \frac{dV}{dr} \times \frac{dr}{dt}$ ← The question asks you to find the rate of increase of the radius — i.e. $\frac{dr}{dt}$.

Using the chain rule: $5 = 4\pi r^2 \times \frac{dr}{dt}$

$5 = 61.18 \times \frac{dr}{dt}$ ← Using r = 2.21 cm

$\frac{dr}{dt} = 5 \div 61.18 = 0.082 \text{ cm/s}$

Example: The area covered by a circular puddle increases at a constant rate of 10 cm²/min. Find the rate of increase in the radius when the radius is 38 cm, to 3 s.f.

The area of the puddle, A, in terms of the radius, r, is given by $A = \pi r^2$.

Differentiate with respect to r: $\frac{dA}{dr} = 2\pi r$

Write out the chain rule: $\frac{dA}{dt} = \frac{dA}{dr} \times \frac{dr}{dt}$ ← The question asks for the rate of increase of the radius — $\frac{dr}{dt}$ again.

Plug in the values: $10 = 2\pi r \times \frac{dr}{dt}$

When r = 38 cm: $\frac{dr}{dt} = \frac{10}{2\pi \times 38} = 0.0419 \text{ cm/min to 3 s.f.}$

Connected Rates of Change

When you've got **Two Different Quantities** — use the chain rule **Twice**

e.g. If you have a question about rates of change of area and volume, you need to consider each one separately.

Example: The surface area of a cube is increasing at the rate of 6 cm²/s.
Find the rate of increase of the volume when the edges are 10 cm long.

The first step is to find the rate of increase of the length — i.e. $\frac{dl}{dt}$

The surface area, A, in terms of the length, l, is given by $A = 6l^2$: $\frac{dA}{dl} = 12l$

Chain rule can be written as $\frac{dA}{dt} = \frac{dA}{dl} \times \frac{dl}{dt}$: $6 = 12l \times \frac{dl}{dt}$ (The question says that $\frac{dA}{dt} = 6$)

when $l = 10$: $6 = 120 \times \frac{dl}{dt} \Longrightarrow \frac{dl}{dt} = 6 \div 120 = \frac{1}{20}$ **cm/s**

Now think about the volume, V: $V = l^3$ so $\frac{dV}{dl} = 3l^2$

Chain rule can be written as $\frac{dV}{dt} = \frac{dV}{dl} \times \frac{dl}{dt}$: $\frac{dV}{dt} = 3l^2 \times \frac{dl}{dt} = 3 \times 10^2 \times \frac{1}{20} = 15$ cm³/s

Remember that **dx/dy = 1 ÷ dy/dx**

There's a handy little trick you can use to differentiate inverse functions. Read on...

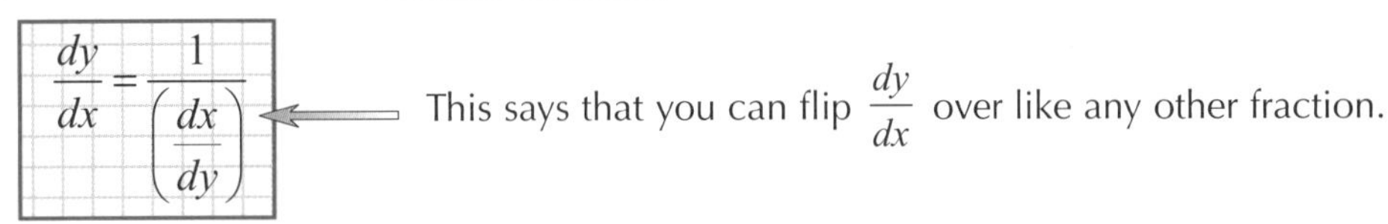

Example: Prove that if y = ln x, then $\frac{dy}{dx} = \frac{1}{x}$. (Assume that differentiating e^x with respect to x gives e^x.)

It asks you to prove the result, so you can't just quote it.
But it gives you a clue — it suggests pretty strongly that you're going to need to differentiate e^x.

It tells you that $y = \ln x$

So take exponentials of both sides to get: $e^y = x$ (or $x = e^y$)

From here, it's easy to find $\frac{dx}{dy}$ — this is just: $\frac{dx}{dy} = e^y$

Flip this over to get $\frac{dy}{dx}$: $\frac{dy}{dx} = \frac{1}{e^y} = \frac{1}{x}$

When you have y in terms of x, you differentiate to find $\frac{dy}{dx}$ — this is differentiation with respect to x. But here, you've got x in terms of y instead, so differentiate with respect to y to get $\frac{dx}{dy}$.

Practice Questions

1) ***Some ink leaks out of a pen in my pocket causing a circular stain which increases in area by 3 cm²/hour. Find the rate of increase of the radius of the stain, when the radius is 2 cm.***
2) ***A spherical beach ball is inflated at a constant rate of 16 cm³/s. Find the rate of increase in the radius when the volume is 3000 cm³.***
3) ***The volume of a sphere increases at a rate of 10 cm³/s. Find the rate of increase of the surface area, when the radius is 8 cm.***

Singalongamaths again — 'Un-Chain-Rule my Heart'...

These can be nasty — it's sometimes tricky to see exactly how to use the Chain Rule, even if you know that's what you're supposed to be using. Try writing down formulas for any quantity mentioned in the question and differentiating them — then see if you can connect them in some way. Or you might be able to get a clue if there are any units mentioned. For example, say an area A changes at a rate given in cm² per second (cm²/s). This rate of change (rate of change = derivative) has units of 'area ÷ time' — so it's a value for dA/dt. This might give you a bit of a clue about how to use the Chain Rule.

Integration

Integration... as if you haven't suffered enough already. No, I'm just joking — it's no harder than differentiation.

Up the power by **One** — then **Divide** by it

The formula below tells you how to integrate any power of x (except x^{-1}).

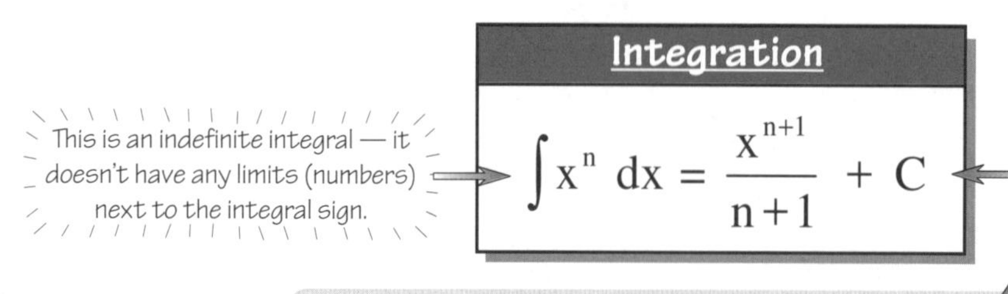

You can't do this to $\frac{1}{x} = x^{-1}$. When you increase the power by 1 (to get zero) and then divide by zero — you get big problems.

In a nutshell, this says:

To integrate a power of x: (i) Increase the power by one — then divide by it.
and (ii) Stick a constant on the end.

Examples: Use the integration formula...

① For 'normal' powers,

$$\int x^3 dx = \frac{x^4}{4} + C$$

Increase the power to 4... ...and then divide by 4.

② For negative powers,

$$\int \frac{1}{x^3} dx = \int x^{-3} dx$$
$$= \frac{x^{-2}}{-2} + C$$
$$= -\frac{1}{2x^2} + C$$

Increase the power by 1 to −2... ...and then divide by −2.

③ For fractional powers,

$$\int \sqrt[3]{x^4}\,dx = \int x^{\frac{4}{3}} dx$$
$$= \frac{x^{\frac{7}{3}}}{(7/3)} + C$$
$$= \frac{3\sqrt[3]{x^7}}{7} + C$$

Add 1 to the power... ...then divide by this new power.

④ And for complicated looking stuff...

$$\int\left(3x^2 - \frac{2}{\sqrt{x}} + \frac{7}{x^2}\right)dx = \int\left(3x^2 - 2x^{-\frac{1}{2}} + 7x^{-2}\right)dx$$
$$= \frac{3x^3}{3} - \frac{2x^{\frac{1}{2}}}{(1/2)} + \frac{7x^{-1}}{-1} + C$$
$$= x^3 - 4\sqrt{x} - \frac{7}{x} + C$$

Do each of these bits separately.

⑤ And if the question has limits, put them in just like you've done before.

$$\int_1^4 \sqrt{x} - \frac{1}{3x^2}\,dx = \int_1^4 x^{\frac{1}{2}} - \frac{1}{3}x^{-2}\,dx$$
$$= \left[\frac{x^{\frac{3}{2}}}{\frac{3}{2}} - \frac{1}{3}\left(\frac{x^{-1}}{-1}\right)\right]_1^4$$
$$= \left[\frac{2}{3}x^{\frac{3}{2}} + \frac{1}{3x}\right]_1^4$$
$$= \left[\left(\frac{2}{3}\times 4^{\frac{3}{2}} + \frac{1}{3\times 4}\right) - \left(\frac{2}{3}\times 1^{\frac{3}{2}} + \frac{1}{3}\right)\right]$$
$$= \left[\left(\frac{16}{3} + \frac{1}{12}\right) - \left(\frac{2}{3} + \frac{1}{3}\right)\right]$$
$$= \frac{53}{12}$$

Change to powers of ½ and -2.

Integrate, put in the limits of 4 and 1, and subtract.

Check your answers:
You can check you've integrated properly by differentiating the answer — you should end up with the thing you started with.

Integration

Integration is the opposite of differentiation. If you differentiate something and then integrate the result, you get back to what you started with (give or take a constant).

Find Curves and Areas by Integrating

Add a constant of integration if your integral doesn't have limits (like when you know a curve's derivative, and you want to find the equation of the curve itself).

Example: Find the equation of the curve which passes through the point (1,5) with $\frac{dy}{dx} = 4\sqrt[3]{x}$

Integrating both sides gives:

$$y = \frac{4x^{4/3}}{4/3} + C \quad \text{so} \quad \mathbf{y = 3x^{4/3} + C}$$

It's easier if you change to $\frac{dy}{dx} = 4x^{1/3}$ before you integrate.

Now find C by putting x = 1 and y = 5 into the equation:

$$5 = 3(1^{4/3}) + C = 3 + C$$
$$\mathbf{C = 2}$$

So the answer is $y = 3x^{4/3} + 2$

But if your integral's got limits (like when you're finding the area under a graph), don't bother with the constant.

Example: Find the areas A and B under the graph below.

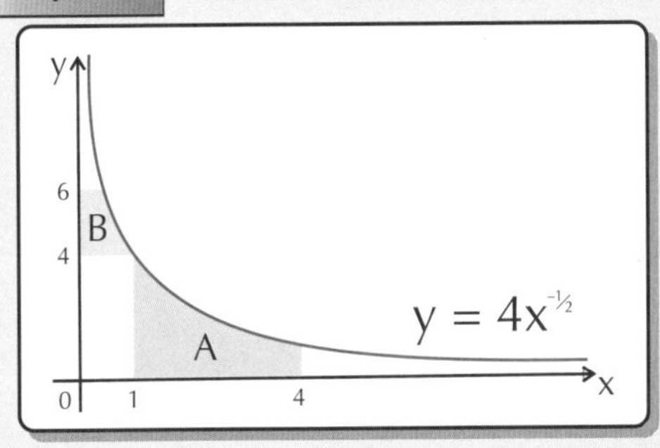

Area A is pretty straightforward:

$$A = \int_1^4 y\,dx = \int_1^4 4x^{-1/2}\,dx$$
$$= \left[\frac{4x^{1/2}}{1/2}\right]_1^4 = \left[8x^{1/2}\right]_1^4$$
$$= \left[(8\times4^{1/2}) - (8\times1^{1/2})\right] = (8\times2) - (8\times1) = \mathbf{8}$$

Area B is a bit nastier — you need to use $B = \int_4^6 x\,dy$

You need to put x in terms of y, so rearrange:

$$y = 4x^{-1/2}$$
$$y^2 = 16x^{-1}$$
$$xy^2 = 16$$

Get x on its own: $x = \frac{16}{y^2}$

$$B = \int_4^6 \frac{16}{y^2}\,dy = \int_4^6 16y^{-2}\,dy$$
$$= \left[\frac{16y^{-1}}{-1}\right]_4^6 = \left[\frac{-16}{y}\right]_4^6$$
$$= \left[\left(-\frac{16}{6}\right) - \left(-\frac{16}{4}\right)\right]$$
$$= -\frac{8}{3} + 4 = \frac{4}{3}$$

Practice Questions

1) Integrate the following: a) $\int 10x^4\,dx$ b) $\int \frac{4}{x^3}\,dx$ c) $\int (3x^3 + 2x^2)\,dx$ d) $\int 4\sqrt[3]{x}\,dx$ e) $\int \left(6x^5 - \frac{2}{x^2} + \sqrt{x}\right)dx$

2) Find the equation of the curve which passes through (1,0) with $\frac{dy}{dx} = \sqrt{x} + \frac{2}{x^2}$

3) Evaluate the following: a) $\int_1^2 \left(\frac{8}{x^5} + \frac{3}{\sqrt{x}}\right)dx$ b) $\int_1^6 \frac{3}{y^2}\,dy$

4) Find area A in the graph on the right:

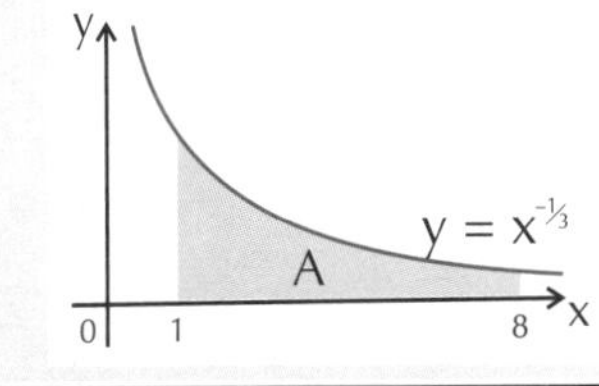

What we hope ever to do with ease, we must learn first to do with diligence...

That's what Dr Johnson reckoned, anyway. And he knew a thing or two, I can tell you. I don't know exactly what he had in mind when he said it, admittedly, but it's true about integration anyway. At first, you'll need all your powers of concentration to avoid making mistakes. But with a bit of practice, it'll become as easy as falling off a log.

Integrating More Functions

Since integration is the reverse of differentiation, it's easy to integrate stuff involving exponentials.

The **Integral** of e^x is e^x

It doesn't come much easier than this...

$$\int e^x dx = e^x + C$$

If there's a number in front you can just 'ignore' it — and if there are limits then just pop them on like before.

$$\int ke^x dx = ke^x + C$$

Example: $\int 2e^x dx = 2e^x + C$

Example:

$$\int_0^1 4e^x dx = [4e^x]_0^1$$
$$= 4e^1 - 4e^0$$
$$= 4e - 4$$
$$= 4(e-1)$$

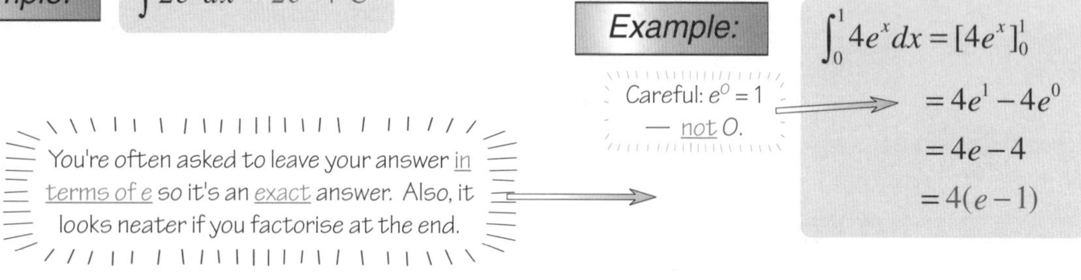

The Integral of $\frac{1}{x}$ is ln x

When you differentiate ln x you get $\frac{1}{x}$. So when you integrate $\frac{1}{x}$, you must get a natural log (ln).

$$\int \frac{1}{x} dx = \ln|x| + C$$

$$\int \frac{k}{x} dx = k\ln|x| + C$$

Don't forget the modulus sign round the x. It's there to cope with negative values (they could create problems when you take logs of them). Don't worry about it — just make sure you include the modulus.

Example: $\int \frac{5}{x} dx = 5\ln|x| + C$

If the question involves limits you might need to use the log laws.

Example:

$$\int_2^7 \frac{5}{x} dx = [5\ln|x|]_2^7$$
$$= 5\ln 7 - 5\ln 2$$
$$= 5(\ln 7 - \ln 2)$$
$$= 5\ln\frac{7}{2}$$

Factorise out the 5 and use the log law $\ln x - \ln y = \ln\frac{x}{y}$

Log Laws

$$\ln(xy) = \ln x + \ln y$$
$$\ln\frac{x}{y} = \ln x - \ln y$$
$$\ln x^n = n\ln x$$

Integrating More Functions

This isn't too hard in itself, but they make it harder in the Exam by including integrals in more difficult topics, or by mixing up a number of functions within a question — like this:

Example:

A curve is given by the equation $y = \sqrt{x} + \frac{12}{x}$ for $x > 0$.

Find the area bounded by the curve in the first quadrant, the lines x = 1, x = 4 and the x-axis. Give your answer in the form $p + q\ln 2$, where p and q are numbers to be found. **[7 marks]**

It all looks pretty complicated — but basically, the word area means you've got to integrate. The limits are hidden away in the question as the x-values 1 and 4.

This is the area you need.

So:

$$Area = \int_1^4 \left(\sqrt{x} + \frac{12}{x}\right)dx$$ [2 marks]

Change the root to a power of $\frac{1}{2}$:

$$= \int_1^4 \left(x^{\frac{1}{2}} + \frac{12}{x}\right)dx$$

Dividing by $\frac{3}{2}$ is the same as multiplying by $\frac{2}{3}$.

Two integrals — the first is just adding one to the power and dividing by the new power. The second integral becomes a log.

$$= \left[\frac{x^{\frac{3}{2}}}{3/2} + 12\ln|x|\right]_1^4 = \left[\frac{2}{3}x^{\frac{3}{2}} + 12\ln|x|\right]_1^4$$ [2 marks]

Plug in the limits 4 and 1, and subtract:

$$= \left[\left(\frac{2}{3}\times 4^{\frac{3}{2}} + 12\ln 4\right) - \left(\frac{2}{3}\times 1^{\frac{3}{2}} + 12\ln 1\right)\right]$$ [2 marks]

Watch out — ln 1 is 0.

$$= \left[\frac{2}{3}\times 8 + 12\ln 4 - \frac{2}{3}\right]$$

The question asked to put the answer in terms of ln 2. Just change ln 4 to ln 2^2 or 2ln 2 (using the log laws):

$$= \frac{14}{3} + 12\ln 4 = \frac{14}{3} + 12\ln 2^2 = \frac{14}{3} + 24\ln 2$$ [1 mark]

Practice Questions

1) *Integrate* $x^{\frac{5}{2}} + \frac{5}{2}e^x$

2) *Integrate* $(x + \frac{1}{x^2})^2$

3) *Integrate* $\sqrt[3]{x} - \frac{3}{2}e^x + \frac{4}{3x}$

4) ***Sample exam question:***

Find the area bounded by the curve $y = 4x - \frac{2}{x}$, the lines x = 1 and x = 3 and the x-axis. [7 marks]

There's no limit to the excitement of integration...

The constant of integration is important — so each time you integrate something, ask yourself if you need one. And don't forget the modulus sign when you integrate 1/x — chances are you won't need to actually use it, but they'll want you to put it there anyway. So learn the basics well, then you can use more brain power on thinking about the hard stuff.

Volume of Revolution

Skip these two pages if you're doing OCR B, AQA A or AQA B.

'You say you want a revolution', sang the Beatles — it's a little-known fact that they were all secret integration fans.

Volumes of Revolution are Integrals with Limits

Rotated 360° about the x-axis

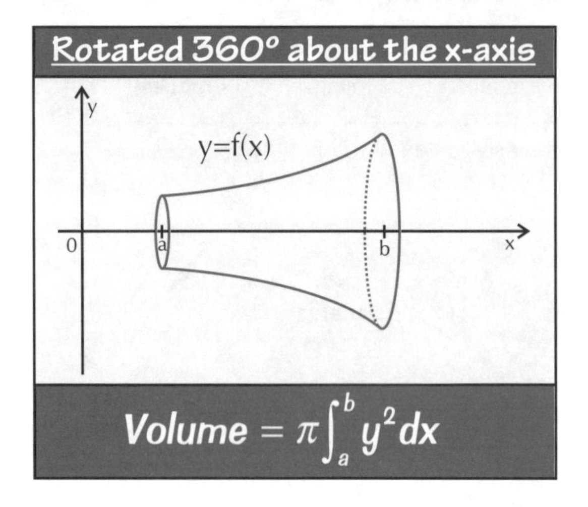

Volume $= \pi \int_a^b y^2 dx$

Rotated 360° about the y-axis

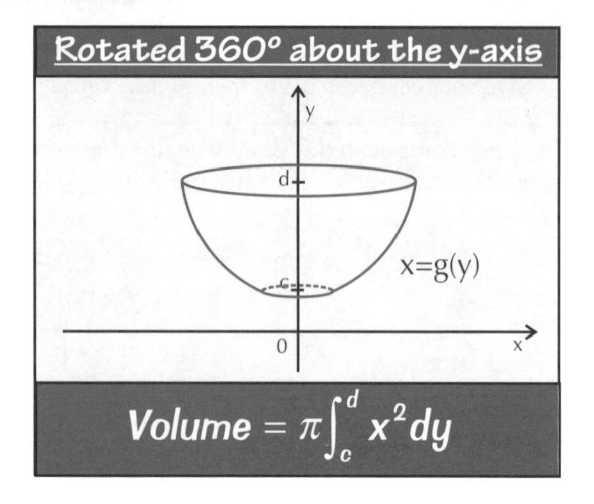

Volume $= \pi \int_c^d x^2 dy$

Example: Find the volume formed when the portion of the curve $y = x^2 + 2$ between x = 1 and x = 3 is rotated 360° about:

a) the x-axis b) the y-axis

a) It's rotating around the x-axis, so the formula is: **volume $= \pi \int y^2 dx$**

Put in the equation of the curve for y and then square it:

$$V = \pi \int_1^3 (x^2+2)^2 dx$$

$$= \pi \int_1^3 (x^2+2)(x^2+2)dx \;\; = \pi \int_1^3 (x^4+4x^2+4)dx$$

Don't forget the middle term.

Integrate, put in the limits 3 and 1 and subtract:

$$= \pi \left[\frac{x^5}{5} + \frac{4x^3}{3} + 4x \right]_1^3$$

$$= \pi \left[\left(\frac{3^5}{5} + \frac{4\times3^3}{3} + 4\times3 \right) - \left(\frac{1^5}{5} + \frac{4\times1^3}{3} + 4 \right) \right]_1^3$$

$$= \pi \left[\left(\frac{243}{5} + \frac{108}{3} + 12 \right) - \left(\frac{1}{5} + \frac{4}{3} + 4 \right) \right] \;\; = \frac{1366\pi}{15}$$

---

b) Now it's rotating around the y-axis, so you need to use the other formula: **volume $= \pi \int x^2 dy$**

The problem is that you need the integral of x^2...

...so rearrange the formula $y = x^2 + 2$ and make x the subject:

$$V = \pi \int_3^{11} x^2 dy \; = \pi \int_3^{11} (y-2)\, dy$$

The limits are different for the y-axis:
When x = 1, y = $1^2 + 2 = 3$
and when x = 3, y = $3^2 + 2 = 11$

Now just complete the integration:

$$= \pi \left[\frac{y^2}{2} - 2y \right]_3^{11}$$

$$= \pi \left[\left(\frac{121}{2} - 22 \right) - \left(\frac{9}{2} - 6 \right) \right] \;\; = 40\pi$$

Volume of Revolution

Volume of revolution Exam Questions are Revolting

OK. Let's see what an exam question looks like — there's every chance they'll slip one onto the paper.

Example:

Find the volume formed when the curve $y = (2x + \frac{1}{x^2})$ is rotated about the x-axis between x = 2 and x = 3. [6 marks]

Use the formula $\textbf{volume} = \pi \int y^2 dx$

Put in $(2x + \frac{1}{x^2})$ and square it:

$$V = \pi \int_2^3 (2x + \frac{1}{x^2})^2 dx = \pi \int_2^3 (4x^2 + \frac{4}{x} + \frac{1}{x^4})\, dx$$ [1 mark]

Change the $\frac{1}{x^4}$ to x^{-4}:

$$= \pi \int_2^3 (4x^2 + \frac{4}{x} + x^{-4}) dx$$

The middle term becomes $4\ln|x|$.

Now integrate the normal way:

$$= \pi \left[\frac{4x^3}{3} + 4\ln|x| - \frac{x^{-3}}{3} \right]_2^3 = \pi \left[\frac{4x^3}{3} + 4\ln|x| - \frac{1}{3x^3} \right]_2^3$$ [2 marks]

Put in the limits and use the log law to tidy up the answer:

$$= \pi \left[\left(\frac{4 \times 3^3}{3} + 4\ln 3 - \frac{1}{3 \times 3^3} \right) - \left(\frac{4 \times 2^3}{3} + 4\ln 2 - \frac{1}{3 \times 2^3} \right) \right]$$ [1 mark]

$$= \pi \left[\left(36 + 4\ln 3 - \frac{1}{81} \right) - \left(\frac{32}{3} + 4\ln 2 - \frac{1}{24} \right) \right]$$ [1 mark]

$$= \pi \left[25\frac{235}{648} + 4\ln\frac{3}{2} \right]$$ [1 mark]

Practice Questions

Sample exam questions:

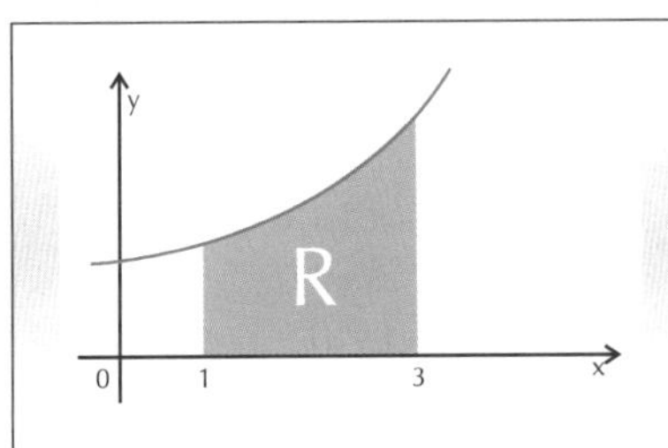

1) The graph shows part of the curve with equation $y = x^2 + 1$. The shaded region R bounded by the curve, the x-axis and the lines x = 1 and x = 3 is rotated through 360° about the x-axis. Find the volume of the solid generated. [6 marks]

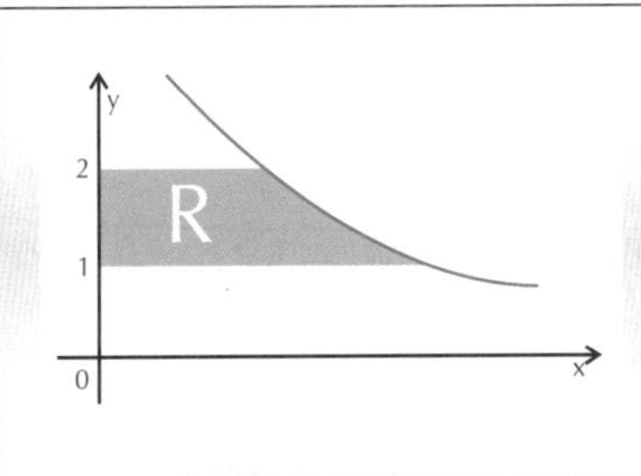

2) The graph shows part of the curve with equation $y = \frac{4}{x^2}$.

The shaded region R bounded by the curve, the y-axis and the lines y = 1 and y = 2 is rotated through 360° about the y-axis. Find the volume of the solid generated. [6 marks]

I'm spinning around, move out of my way...

You've got to use a different formula depending on which axis the curve's being rotated around. If it's going around the x-axis, use the formula ending 'dx'. And if it's going around the y-axis, use the one ending 'dy'. Simple as that. These exam questions tend to cover a lot of topics — volumes, indices, integration, log laws... Better have a lie down.

Integration by Substitution

Skip these two pages if you're doing Edexcel, AQA A or AQA B.

This sounds more complicated than it really is. If you've learnt the other rules for integration you should be okay.

Substitution is all about Replacing

Substitution involves replacing x with a linear term — i.e. something of the form ax + b, where a and b are numbers. Then you need to pretend 'ax + b' is really just an x, integrate as normal, and then divide by a at the end.

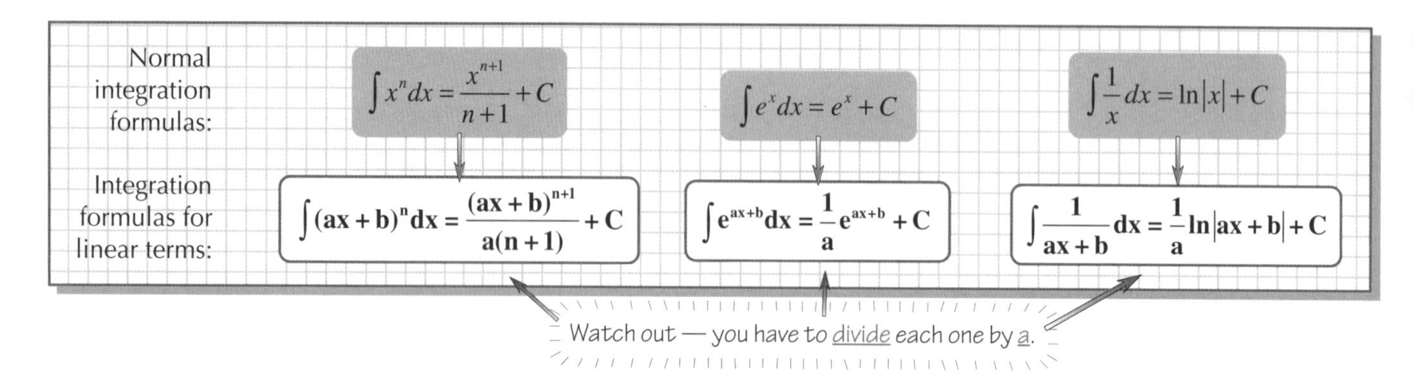

Example: Solve $\int (2x-1)^8 dx$

The first step is the same as normal — up the power by 1 to 9, then divide by 9. Then you also need to divide by 2 (the **a** from the bracket). This gives:

$$\int (2x-1)^8 dx = \frac{(2x-1)^9}{2\times 9} + C = \frac{1}{18}(2x-1)^9 + C$$

Example: Solve $\int e^{3x+2} dx$

The exponential bit stays the same and you just divide by a = 3:

$$\int e^{3x+2} dx = \frac{1}{3} e^{3x+2} + C$$

Example: Solve $\int \frac{1}{4x-5} dx$

This one's going to turn into a log — and don't forget to divide by 4.

$$\int \frac{1}{4x-5} dx = \frac{1}{4} \ln|4x-5| + C$$

Example: Show that $\int_0^1 \frac{3}{2x+5} dx = \frac{3}{2} \ln\left(\frac{7}{5}\right)$

Two extra problems for you here — the 3 on the top and the limits. Just multiply the answer by 3 and put in the limits as usual.

Using the log law $\ln\left(\frac{a}{b}\right) = \ln a - \ln b$

$$\int_0^1 \frac{3}{2x+5} dx = \left[3 \times \frac{1}{2} \ln|2x+5|\right]_0^1$$

$$= \left[\frac{3}{2} \ln|2x+5|\right]_0^1 = \frac{3}{2}\ln 7 - \frac{3}{2}\ln 5 = \frac{3}{2}(\ln 7 - \ln 5) = \frac{3}{2}\ln\left(\frac{7}{5}\right)$$

Integration by Substitution

So far so good, but there's still a few things the examiners can try to catch you out with. A few fractions (plus some negative numbers thrown in for good measure) are the next challenge. Just keep your head and apply the normal rules.

Integration by Substitution Applies to Other Topics

This is the trickiest kind of integration, so it's a fair bet it'll turn up in the Exam.

Example: Find the area enclosed by the curve $y = \sqrt{8x+1}$, the x-axis and x = 0 and x = 1.

$$\text{Area} = \int_0^1 \sqrt{8x+1}\,dx = \int_0^1 (8x+1)^{\frac{1}{2}}\,dx$$

Start by changing the $\sqrt{\ }$ to a power of ½.

You need to up the power to $\frac{3}{2}$...

...then divide by $\frac{3}{2}$ and 8:

$$\left[\frac{(8x+1)^{\frac{3}{2}}}{8\times\frac{3}{2}}\right]_0^1 = \left[\frac{1}{12}(8x+1)^{\frac{3}{2}}\right]_0^1$$

$$= \left[\frac{1}{12}\times 9^{\frac{3}{2}} - \frac{1}{12}\times 1^{\frac{3}{2}}\right]$$

$$= \frac{27}{12} - \frac{1}{12}$$

$$= \frac{13}{6}$$

Example: Find the equation of the curve which has gradient e^{-x} and passes through the point (ln 2, 1).

The gradient $\frac{dy}{dx} = e^{-x}$ so to find y you need to integrate.

e^{-x} stays as e^{-x} but you need to divide by -1: $y = -e^{-x} + C$

To find C you now put in x = ln 2 and y = 1: $1 = -e^{-\ln 2} + C$

Use the log laws: $-e^{-\ln 2} = -e^{\ln 2^{-1}} = -e^{\ln(\frac{1}{2})}$

$$= -e^{\ln 2^{-1}} + C = -e^{\ln(\frac{1}{2})} + C$$

$$= -\tfrac{1}{2} + C$$

The e and ln cancel.

so $1 = -\frac{1}{2} + C$ which means $C = \frac{3}{2}$. The equation of the curve is $y = -e^{-x} + \frac{3}{2}$

Practice Questions

1) *Integrate these:* a) $(7x + 2)^3$ b) $\frac{1}{(2x-3)^2}$ c) $\frac{1}{(2x-3)}$ d) $2e^{5x-1}$ e) $\frac{3}{4x+1}$ f) $\frac{9}{(3x-2)^4}$

2) *Show that* $\int_2^3 \frac{5}{2x-1}\,dx = \frac{5}{2}\ln\left(\frac{5}{3}\right)$

Sample exam questions:

3) Find the equation of the curve which has gradient $\frac{2}{4x-3}$ and passes through the point (1,3). [5 marks]

4) If $y = \left(1 + \frac{1}{\sqrt{2x+1}}\right)$ find $\int y^2\,dx$ [5 marks]

The true art of memory is the art of attention...

Yep — it's Dr Johnson again. But he's right — you'll never remember this if you don't concentrate. Since integration is the opposite of differentiation, you can always check you've integrated properly by differentiating and seeing that you get back to what you started with (no cheating allowed). Use the Chain Rule (see p44) to differentiate your answer if you need to.

Change of Sign

Skip these two pages if you're doing AQA B.

This is all about finding solutions or roots of an equation. It can involve lots of calculator-button pressing.

Finding a ***Change in Sign*** *Helps You* ***Locate*** *a* ***Root***

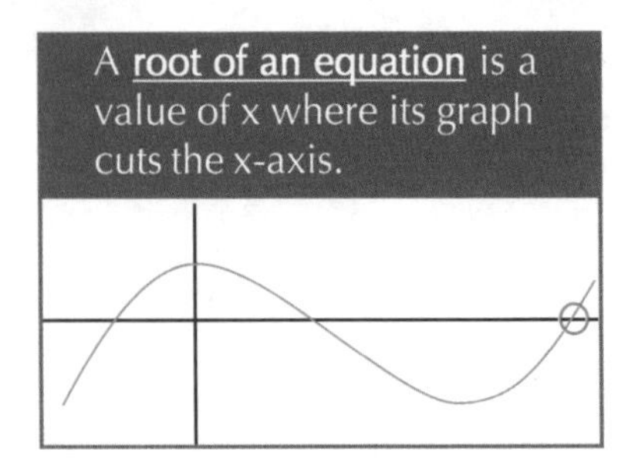

The point where the graph hits the x-axis is a root.

You can see that to the left of the root the graph is below the x-axis — so it's negative. To the right it's above the x-axis — so it's positive.

That means that you can just look for a change of sign, and that'll show whereabouts the root lies.

If you plug two numbers into the equation of the curve and there's a change of sign (i.e. one of the values is positive and the other is negative), then the root must be between those two numbers.

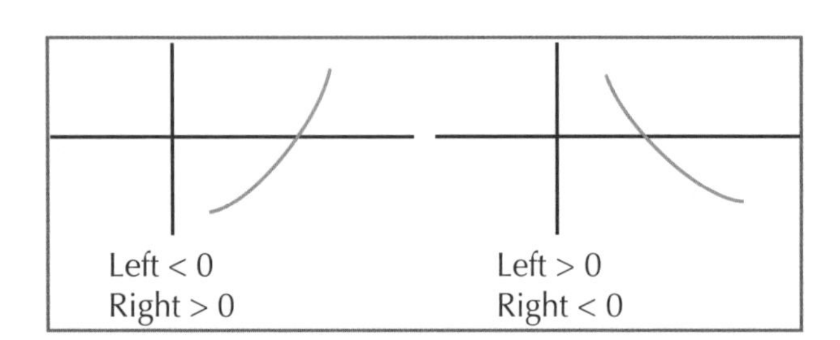

Example: Show there is a root of the equation $x^2 = 5x - 2$ in the interval [4,5].

Start by collecting all the terms on the left so the equation equals 0: $x^2 - 5x + 2 = 0$

Then plug 4 into the equation: $4^2 - 5 \times 4 + 2 = 16 - 20 + 2 = -2$ (which is less than 0)

Next plug 5 into the equation: $5^2 - 5 \times 5 + 2 = 25 - 25 + 2 = 2$ (which is greater than 0)

So there's a change of sign and the root must lie in the interval [4,5].

Check the ***Accuracy*** *of* ***Approximations*** *using* ***Change of Sign***

This is pretty much the same as above. If you can show that the root is between the minimum and maximum values possible for an approximation — bingo. You've proved it's accurate.

Example: 6.937 is an approximation to a root of the equation $\ln x - x + 5 = 0$. Check its accuracy to 3 decimal places.

If 6.937 is correct to 3 decimal places then the root must lie in the interval [6.9365, 6.9375).
Think about it — anything between these two values will round to 6.937.
Using your calculator and putting 6.9365 and 6.9375 into the equation gives:

ln(6.9365) – 6.9365 + 5 =0.000297 (which is greater than 0)

ln(6.9375) – 6.9375 + 5 = -0.000559 (which is less than 0)

This shows a change of sign and so the root is 6.937 to 3 decimal places.

Change of Sign

Always Draw a Sketch for Change of Sign questions

You know it makes sense — draw a sketch even if you're not sure what the graph really looks like. It will help. Honest.

Example:

The diagram shows part of the curve $y = f(x)$ where $f(x) = 8 + \ln 2x - e^x$ and $0.1 \le x \le 4.0$

Given that $f(a) = 0$ show that $2.2 \le a \le 2.3$.

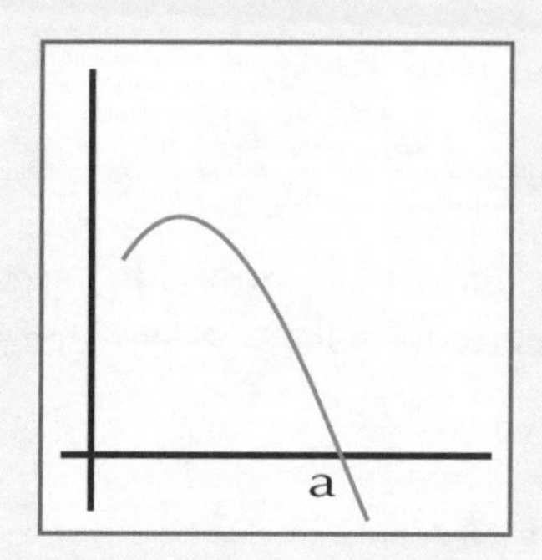

The curve cuts the x-axis at a (because $f(a) = 0$).
There are no real complications here — the question is just asking to check for a root in the interval [2.2, 2.3].

First put in 2.2: so $f(2.2) = 8 + \ln(2 \times 2.2) - e^{2.2} = 0.456...$
— which is greater than 0.

Putting in 2.3 gives $f(2.3) = 8 + \ln(2 \times 2.3) - e^{2.3} = -0.448...$
— which is less than 0.

An obvious change of sign, so the root a lies in the interval [2.2, 2.3].

Practice Questions

1) ***Show that a root of the equation $x^3 - 5x - 4 = 0$ lies in the interval [2, 3].***

2) ***4.32 is an approximation to a root of the equation $x \ln x - 2 - x = 0$. Check its accuracy to 2 decimal places.***

3) ***Sample exam question:***

a) Show by means of a sketch that the equation $2 - x^2 = e^x$ has 2 roots. [2 marks]

b) Show that one of the roots lies in the interval [0, 1]. [2 marks]

c) The other root lies in the interval [n, n+1] where n is an integer. Find n. [2 marks]

Maths rhyming slang #1: Fog on the Tyne — change of sign...

Its not that bad really, just put two values into an equation, check for a change of sign and the job's done. Well, almost done — you still have to write Change of Sign — and you'll lose marks if you don't. Think of it as just another hoop you have to jump through — it may not make much sense, but the examiners will like you more if you do.

Iteration

Skip these two pages if you're doing AQA A or AQA B.

It's... errrrrr... it's all about knowing where to start, it's all about perseverance, and it's all about knowing when to stop.

Solving Equations Using Iteration is Repetitive

To solve an equation to a certain degree of accuracy, you need an iterative formula. It'll be in this form:

$$x_{n+1} = g(x_n)$$

...plus a starting x value — usually x_0.

Example: $x_{n+1} = \sqrt{2x_n + 1}$ with $x_0 = 1$

The $g(x_n)$ here is $\sqrt{2x_n + 1}$ and the starting value (the first 'guess' of the answer to the equation) is 1.
Start with x_0 and put it into the right-hand side of the iterative formula. The left-hand side will be x_1.

So $x_1 = \sqrt{2x_0 + 1} = \sqrt{2 \times 1 + 1} = \sqrt{3} = 1.732050808$

Now put this x_1 into the right-hand side of the formula to get x_2:

$x_2 = \sqrt{2x_1 + 1} = \sqrt{2 \times 1.732050808 + 1} = \sqrt{4.464101615} = 2.112842071$

And this plugs in to get x_3:

$x_3 = \sqrt{2x_2 + 1} = \sqrt{2 \times 2.112842071 + 1} = \sqrt{5.225684141} = 2.285975534$

This process continues giving $x_4 = 2.360498055$, $x_5 = 2.391860387$, $x_6 = 2.40493675$,
$x_7 = 2.410367918$ ... $x_{11} = 2.414100295$, $x_{12} = 2.414166645$

The ANS Button on your calculator can Help

The ANS (or 'answer') button remembers the last answer you got, so you don't have to type out the number.

So here's that last example again:

Put 1 in your calculator and press [EXE] or [=]

Enter $\sqrt{\ }(2 \times ANS + 1)$ and press [EXE] or [=]

Out comes x_1 and if you keep pressing [EXE] you get the rest of the x-values.
Don't forget to write down the values as you get them or you will lose a truckload of method marks.

The Iterative Formula is a Rearrangement of the Original Equation

You need to make x the subject and tag on a few n's to make it an iterative formula.

Example: Find an iterative formula for the equation $10x - x^3 + 7 = 0$ in the form $x_{n+1} = \dfrac{(x_n^3 + p)}{q}$ where p and q are constants to be found.

You need to rearrange the formula to make x the subject...

...so move the x^3 and 7 onto the right: $10x = x^3 - 7$

and divide by 10: $x = \dfrac{x^3 - 7}{10}$

And the iterative formula is: $x_{n+1} = \dfrac{(x_n^3 - 7)}{10}$

So $p = -7$ and $q = 10$.

Iteration

Skip 'Newton-Raphson' if you're doing Edexcel, AQA A, AQA B or OCR A.

The **Newton-Raphson** Method is used to help solve **f(x) = 0**

The Newton-Raphson method is another iterative method. You can use it to solve equations of the form $f(x) = 0$.

If x_n is a rough solution of the equation $f(x) = 0$, then a better solution is given by:
$$x_{n+1} = x_n - \frac{f(x_n)}{f'(x_n)}$$

Example: If $x = 3$ is an approximation to the solution of $x^3 + 3x^2 - 7x - 25 = 0$, use the Newton-Raphson method to find a better solution.

1. You've got a function that you need to make equal to zero — call it $f(x)$. So $f(x) = x^3 + 3x^2 - 7x - 25$
2. Differentiate f(x) to get: $f'(x) = 3x^2 + 6x - 7$
3. Write out the Newton-Raphson formula: $x_{n+1} = x_n - \frac{x^3+3x^2-7x-25}{3x^2+6x-7}$
4. You have a rough solution — call it x_0. So $x_0 = 3$. Now put this in the Newton-Raphson formula to get a more accurate approximation, x_1:
$$x_1 = x_0 - \frac{f(x_0)}{f'(x_0)} = 3 - \frac{f(3)}{f'(3)}$$
$$= 3 - \frac{27+27-21-25}{27+18-7} = 3 - \frac{8}{38} = 2.78947$$
5. Then put x_1 in the Newton-Raphson formula to get an even better approximation, x_2:
$$x_2 = x_1 - \frac{f(x_1)}{f'(x_1)} = 2.78947 - \frac{f(2.78947)}{f'(2.78947)}$$
$$= 2.78947 - \frac{0.52240}{33.08025} = 2.77368$$
6. The more iterations you do, the more accurate your answer: $x_3 = 2.77368 - \frac{0.00290}{32.72198} = 2.77359$

Stop when you get an answer that looks good enough: $2.77359^3 + 3\times2.77359^2 - 7\times2.77359 - 25 = -0.000048$

That seems pretty good — so stop there. (The question will usually tell you when you can stop.)

Practice Questions

1) *Show $x^3 - 2x - 3 = 0$ can be rearranged into the form $x = \sqrt[3]{(2x+3)}$.*
Use the iterative formula $x_{n+1} = \sqrt[3]{(2x_n+3)}$, starting with $x_0 = 2$, to find a root of $x^3 - 2x - 3 = 0$ to 3 d.p.

2) *Show $x^2 - 5x + 2 = 0$ has a root lying in the interval (0, 1) and the equation can be rearranged into the form $x = \frac{(x^2+p)}{q}$ where p and q are constants.*
Using the iterative formula $x_{n+1} = \frac{(x_n^2+p)}{q}$ with your values of p and q, and starting with $x_0 = 0.5$, find one root of the equation $x^2 - 5x + 2 = 0$ to 3 decimal places.

3) *Use two iterations of the Newton-Raphson method to find a solution to the equation $f(x) = 0$, where $f(x) = x^4 - 4x^3 + 2x^2 - 14x + 10 = 0$ and the initial approximation is $x_0 = 1$.*

Maths rhyming slang #2: All sweat and perspiration — iteration...

Not all iterative formulas will converge to a root — the sequence might diverge instead, in which case it's not going to help you much. If you look round in the Exam, you'll be able to tell who's doing the iteration question — it's the people jabbing away at the buttons on the calculator like there's no tomorrow, and quietly cursing now and then as they hit the wrong one.

Numerical Integration of Functions

Skip these two pages if you're doing AQA A, AQA B or OCR B.

Sometimes integrals can be just too hard to do using the normal methods — then you need to know other ways to solve them. That's where the Trapezium Rule comes in.

*The **Trapezium Rule** is Used to Find the **Approximate Area** Under a Curve*

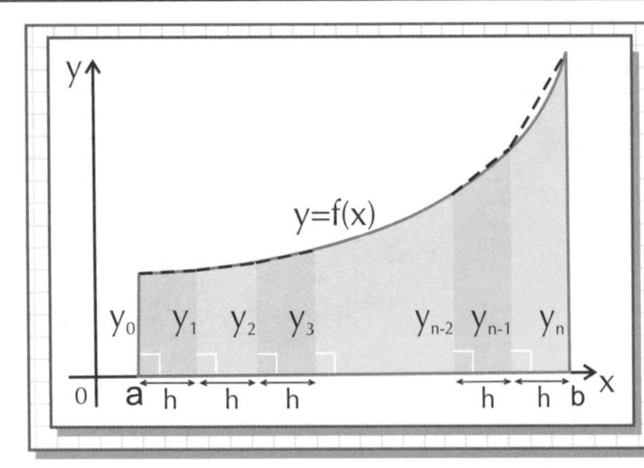

The area represented by $\int_a^b y\,dx$ is approximately:

$$\int_a^b y\,dx \approx \frac{h}{2}[y_0 + 2(y_1 + y_2 + ... + y_{n-1}) + y_n]$$

where **n** is the number of strips or intervals and **h** is the width of each strip.

You can find the width of each strip using $h = \frac{(b-a)}{n}$

$y_0, y_1, y_2, \dots, y_n$ are the heights of the sides of the trapeziums — you get these by putting the x-values into the curve.

So basically the formula for approximating $\int_a^b y\,dx$ works like this:

'Add the first and last heights $(y_0 + y_n)$ and add this to twice all the other heights added up — then multiply by $\frac{h}{2}$.'

Example: Find an approximate value to $\int_0^2 \sqrt{4-x^2}\,dx$ using 4 strips. Give your answer to 4 s.f.

Start by working out the width of each strip: $h = \frac{(b-a)}{n} = \frac{(2-0)}{4} = 0.5$

This means the x-values are $x_0 = 0$, $x_1 = 0.5$, $x_2 = 1$, $x_3 = 1.5$ and $x_4 = 2$ (the question specifies 4 strips, so n = 4).
Set up a table and work out the y-values or heights using the equation in the integral.

x	$y = \sqrt{4-x^2}$
$x_0 = 0$	$y_0 = \sqrt{4-0^2} = 2$
$x_1 = 0.5$	$y_1 = \sqrt{4-0.5^2} = \sqrt{3.75} = 1.936491673$
$x_2 = 1.0$	$y_2 = \sqrt{4-1.0^2} = \sqrt{3} = 1.732050808$
$x_3 = 1.5$	$y_3 = \sqrt{4-1.5^2} = \sqrt{1.75} = 1.322875656$
$x_4 = 2.0$	$y_4 = \sqrt{4-2.0^2} = 0$

Now put all the y-values into the formula with h and n:

$$\int_a^b y\,dx \approx \frac{0.5}{2}[2 + 2(1.936491673 + 1.732050808 + 1.322875656) + 0]$$
$$\approx 0.25[2 + 2 \times 4.991418137]$$
$$\approx 2.996 \text{ to 4 s.f.}$$

Watch out — if they ask you to work out a question with 5 x-values (or 'ordinates') then this is the same as 4 strips. The x-values usually go up in nice jumps — if they don't then check your calculations carefully.

*The Approximation might be an **Overestimate** or an **Underestimate***

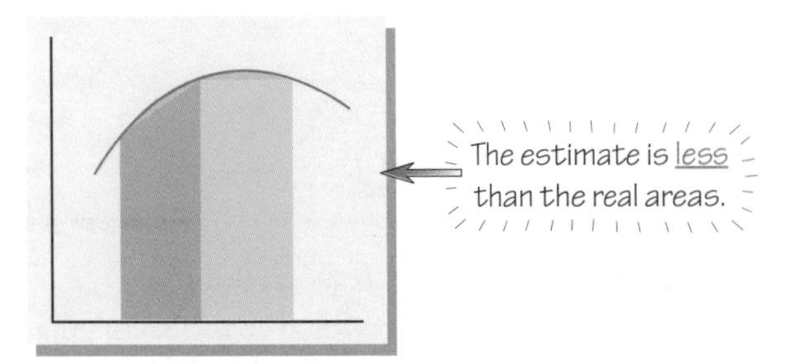

The estimate is less than the real areas.

The estimate is more than the real areas.

Numerical Integration of Functions

These are usually popular questions with examiners — as long as you're careful there are plenty of marks to be had.

The **Trapezium Rule** is in the **Formula Booklet**

...so don't try any heroics — always look it up and use it with these questions.

Example: Use the trapezium rule with 7 ordinates to find an approximation to $\int_1^{2.2} 2\ln x\, dx$

Remember, 7 ordinates means 6 strips — so n = 6.

Calculate the width of the strips: $h = \frac{(b-a)}{n} = \frac{(2.2-1)}{6} = 0.2$

Set up a table and work out the y-values using y = 2ln x:

x	$y = 2\ln x$
$x_0=1.0$	$y_0 = 2\ln 1 = 0$
$x_1=1.2$	$y_1 = 2\ln 1.2 = 0.36464$
$x_2=1.4$	$y_2 = 0.67294$
$x_3=1.6$	$y_3 = 0.94001$
$x_4=1.8$	$y_4 = 1.17557$
$x_5=2.0$	$y_5 = 1.38629$
$x_6=2.2$	$y_6 = 1.57691$

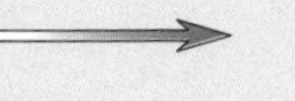

Putting all these values in the formula gives:

$$\int_a^b y\,dx \approx \frac{0.2}{2}[0 + 2(0.36464 + 0.67294 + 0.94001 + 1.17557 + 1.38629) + 1.57691]$$
$$\approx 0.1[1.57691 + 2 \times 4.53945]$$
$$\approx 1.06558$$
$$\approx 1.066 \text{ to 3 d.p.}$$

Practice Questions

1) ***Use the trapezium rule with n intervals to estimate:***

a) $\int_0^3 (9 - x^2)^{\frac{1}{2}} dx$ with $n = 3$

b) $\int_{-1}^{1.5} e^{x^2} dx$; n=5

2) ***Sample exam question:***

The following is a table of values for $y = \ln(2^x)$

x	*0*	*0.5*	*1*	*1.5*	*2*
y	*0*	*0.3466*	*p*	*1.0397*	*q*

a) Find the values of p and q. [2 marks]

b) Use the trapezium rule and all the values of y in the completed table to obtain an estimate of I to 3 decimal places where $I = \int_0^2 \ln(2^x)dx$ [4 marks]

Maths rhyming slang #3: Dribble and drool — Trapezium rule...

Take your time with Trapezium Rule questions — it's so easy to make a mistake with all those numbers flying around. Make a nice table showing all your ordinates (careful — this is always one more than the number of strips). Then add up y_1 to y_{n-1} and multiply the answer by 2. Add on y_0 and y_n. Finally, multiply what you've got so far by the width of a strip and divide by 2. It's a good idea to write down what you get after each stage, by the way — then if you press the wrong button (easily done) you'll be able to pick up from where you went wrong. They're not hard — just fiddly.

Arc Length and Sector Area

Skip these two pages if you're doing Edexcel or OCR B.

Arc lengths and sector areas are easier than you'd think — once you've learnt two simple(ish) formulas.

Always work in **Radians** for **Arc Length** and **Sector Area Questions**

Remember it — for arc length and sector area questions you've got to measure all the angles in radians. The main thing is that you know how radians relate to degrees.

Converting angles

Radians to degrees:	Degrees to radians:
Divide by π, multiply by 180.	Divide by 180, multiply by π.

Here's a table of some of the common angles you're going to need:

Degrees	0	30	45	60	90	120	180	270	360
Radians	0	$\frac{\pi}{6}$	$\frac{\pi}{4}$	$\frac{\pi}{3}$	$\frac{\pi}{2}$	$\frac{2\pi}{3}$	π	$\frac{3\pi}{2}$	2π

If you have part of a circle, like a section of pie chart, you can work out the length of the curved side, or the area of the 'slice of pie' – as long as you know the angle at the centre (θ) and the length of the radius (r).

Arc length and sector area

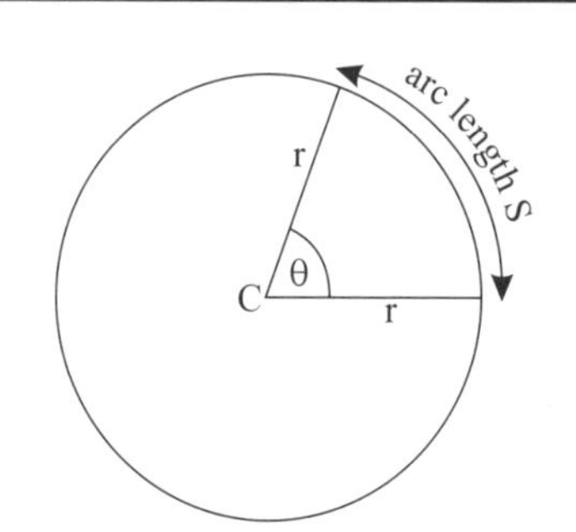

For a circle with a radius of r, where the angle θ is measured in radians, the arc length of the sector S is given by:

$$S = r\theta$$

Put $\theta = 2\pi$ into S and you get: $\mathbf{S = r \times 2\pi = 2\pi r}$.
That's the formula for the circumference of a circle.

You can work out A, the area of the sector, using:

$$A = \tfrac{1}{2} r^2 \theta$$

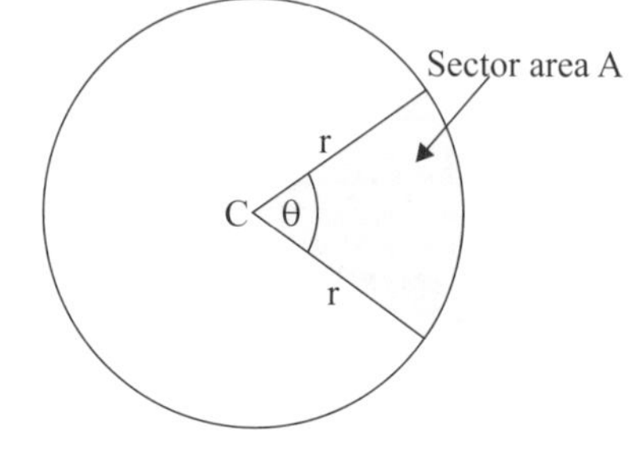

Put $\theta = 2\pi$ into A and you get: $\mathbf{A = \frac{1}{2} r^2 \times 2\pi = \pi r^2}$.
That's the formula for the area of a circle.

You need to tinker around to find **Areas of Segments**

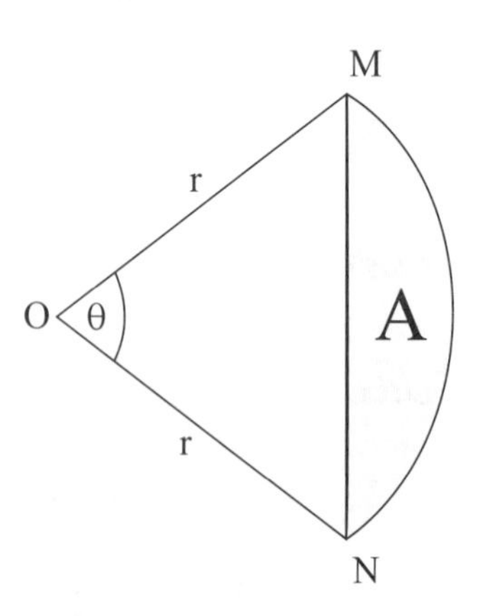

A segment is the 'top bit' of a sector — the yellow bit. You can find its area by subtracting the area of the triangle OMN from the area of the sector.

Area of the whole sector: $A_{sector} = \frac{1}{2} r^2 \theta$

Area of the triangle OMN: $A_{triangle} = \frac{1}{2} r^2 \sin\theta$

So the formula for area A must be: $A = A_{sector} - A_{triangle}$
$= \frac{1}{2} r^2 \theta - \frac{1}{2} r^2 \sin\theta$
$\mathbf{= \frac{1}{2} r^2 (\theta - \sin\theta)}$

Area of a triangle is $A = \frac{1}{2} ab\sin\theta$, where a and b are the lengths of two of the sides, and θ is the angle between them. Here, a and b both equal r.

Arc Length and Sector Area

Questions on trigonometry quite often use the same angles — so it makes life easier if you know the sin, cos and tan of these commonly used angles. Or to put it another way, examiners expect you to know them — so learn them.

Draw Triangles to remember sin, cos and tan of the Important Angles

You need to know the values of sin, cos and tan of $\frac{\pi}{6}$, $\frac{\pi}{3}$ and $\frac{\pi}{4}$ (that's 30°, 60° and 45°).
These two triangles will really help, so draw a quick sketch whenever you've got trig questions in radians.

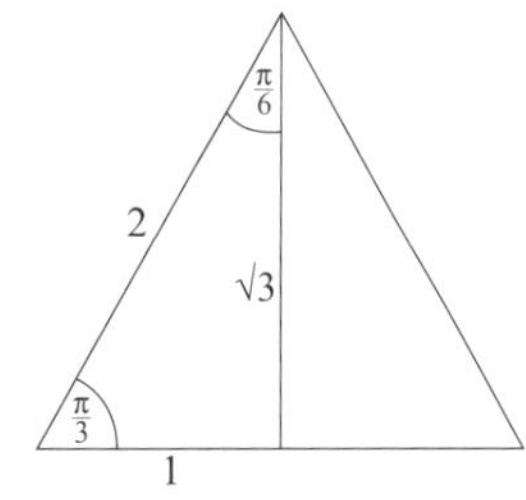

Draw half an equilateral triangle with sides 2 to get sin, cos and tan of $\frac{\pi}{6}$ and $\frac{\pi}{3}$.

Get the height $\sqrt{3}$ by Pythagoras' theorem: $1^2 + \sqrt{3}^2 = 2^2$

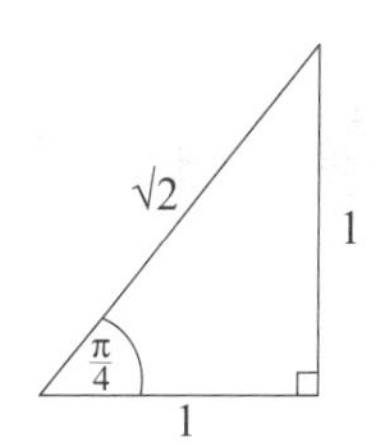

Draw a right-angled triangle with two sides of length 1 to get sin, cos and tan of $\frac{\pi}{4}$.

$\sqrt{2}$ comes from Pythagoras.

You just need to use that old chestnut SOH CAH TOA to work out the values using the triangles:

θ	$\frac{\pi}{6}$	$\frac{\pi}{4}$	$\frac{\pi}{3}$
sinθ	$\frac{1}{2}$	$\frac{1}{\sqrt{2}}$	$\frac{\sqrt{3}}{2}$
cosθ	$\frac{\sqrt{3}}{2}$	$\frac{1}{\sqrt{2}}$	$\frac{1}{2}$
tanθ	$\frac{1}{\sqrt{3}}$	1	$\sqrt{3}$

$\sin = \frac{\text{opp}}{\text{hyp}}$ $\cos = \frac{\text{adj}}{\text{hyp}}$ $\tan = \frac{\text{opp}}{\text{adj}}$

e.g. $\sin\frac{\pi}{4} = \frac{\text{opp}}{\text{hyp}} = \frac{1}{\sqrt{2}}$

Example: Find the area of the shaded part of the symbol.

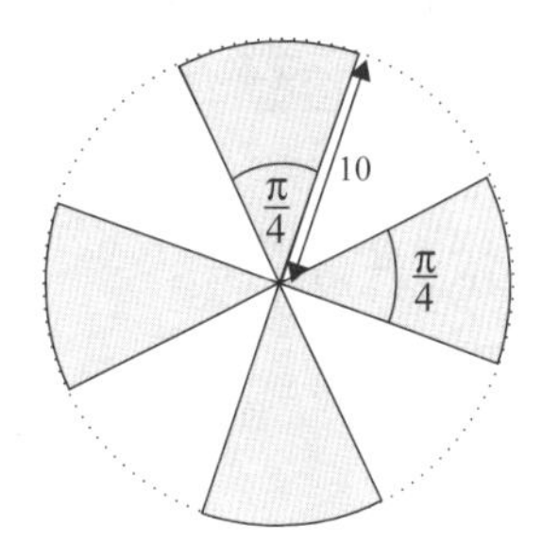

You need the area of the 'leaves' and so use the formula $\frac{1}{2}r^2\theta$.

Each leaf has area $\frac{1}{2} \times 10^2 \times \frac{\pi}{4} = 25\frac{\pi}{2}$ cm²

So the area of the whole symbol $= 4 \times 25\frac{\pi}{2} = 50\pi$ cm²

Example: XYZ is an equilateral triangle. Find the exact area of the shaded segment.

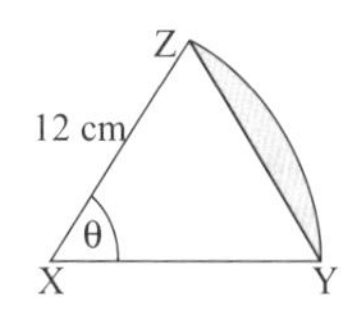

The area of the segment is $\frac{1}{2}r^2(\theta - \sin\theta)$ and here $\theta = 60°$ or $\frac{\pi}{3}$ and the radius r = 12.

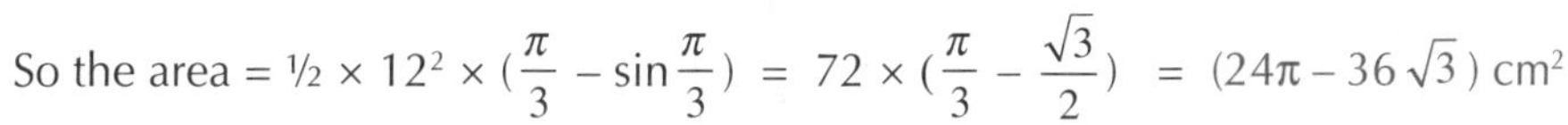

So the area $= \frac{1}{2} \times 12^2 \times (\frac{\pi}{3} - \sin\frac{\pi}{3}) = 72 \times (\frac{\pi}{3} - \frac{\sqrt{3}}{2}) = (24\pi - 36\sqrt{3})$ cm²

Practice Questions

1) ***Find L and A in the diagram:*** (sector with angle 45°, radius 20 cm, arc L, area A)

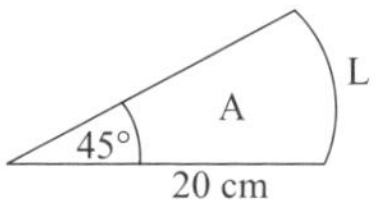

2) ***The diagram shows a sector of radius 8 cm. The length of arc CD is 7.56 cm. Find the angle θ and the area of the sector, A.***

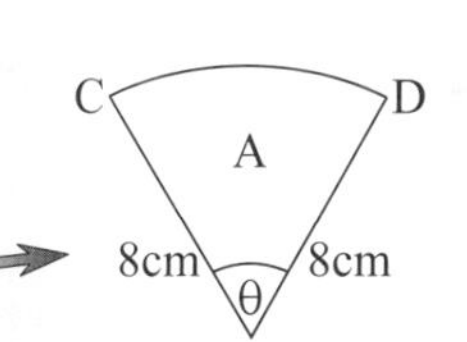

3) ***Sample exam question:***

Find the exact area of the shaded region in the diagram. [6 marks]

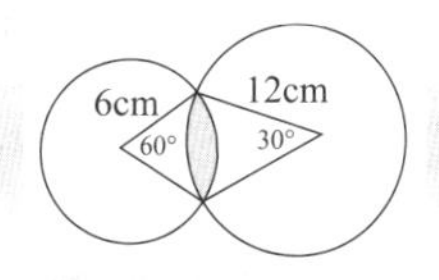

π = 3.14159265358979323846264338327950288419716939... *(Make sure you know it)*

It's worth repeating, just to make sure — those formulas for arc length and sector area only work if the angle is in radians.

Solving Trig Equations in a Given Interval

Skip these two pages if you're doing Edexcel, OCR A or OCR B.

Solving trig equations will no doubt play a very important part in your life from now on. So important, in fact, that you'll probably end up wondering how you managed without it. Yep, the rest of your life starts right here...

There are **Two Ways** to find Solutions in an **Interval...**

Example: Solve $\cos x = \frac{1}{2}$ for $-2\pi \le x \le 4\pi$

Like I said — there are two ways to solve this kind of question. Just use the one you prefer...

You can draw a **graph...**

Your calculator gives you a solution of 60° — convert this into $\frac{\pi}{3}$ using the tricks on p62. Then you have to work out what the others will be.

The other solutions are $\frac{\pi}{3}$ either side of the graph's peaks.

1) Draw the graph of $y = \cos x$ for the range you're interested in...
2) Get the first solution from your calculator and mark this on the graph,
3) Use the symmetry of the graph to work out what the other solutions are:

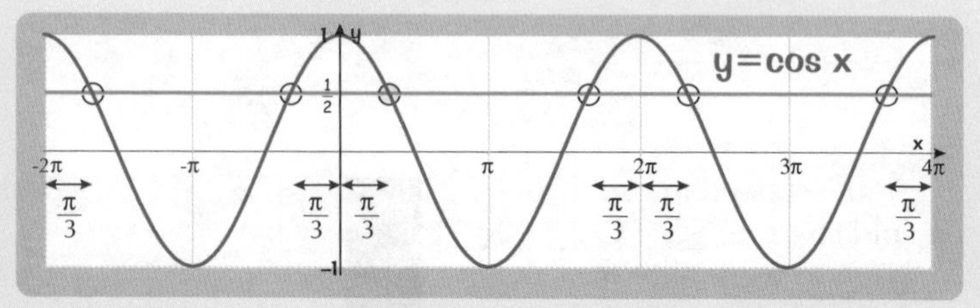

The solutions are: $x = \frac{-5\pi}{3}, \frac{-\pi}{3}, \frac{\pi}{3}, \frac{5\pi}{3}, \frac{7\pi}{3}$ and $\frac{11\pi}{3}$.

...or you can use the **CAST** diagram

CAST stands for COS, ALL, SIN, TAN — and the CAST diagram shows you where these functions are positive:

Between $\frac{\pi}{2}$ and π, only SIN is positive.

Between 0 and $\frac{\pi}{2}$, ALL of sin, cos and tan are positive.

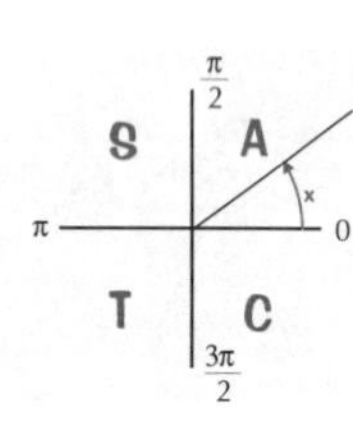

Between π and $\frac{3\pi}{2}$, only TAN is positive.

Between $\frac{3\pi}{2}$ and 2π, only COS is positive.

This is positive — so you're only interested in where cos is positive.

First, to find all the values of x between 0° and 360° where $\cos x = \frac{1}{2}$ — you do this:

Put the first solution onto the CAST diagram.

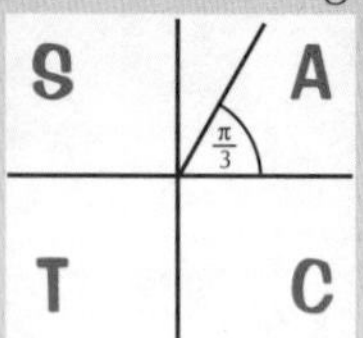

The angle from your calculator goes anticlockwise from the x-axis (unless it's negative — then it would go clockwise into the 4th quadrant).

Find the other angles between 0 and 2π that might be solutions.

The other solutions come from making the same angle from the horizontal axis into the other 3 quadrants.

Ditch the ones that are the wrong sign.

$\cos x = \frac{1}{2}$, which is positive. The CAST diagram tells you cos is positive in the 4th quadrant — but not the 2nd or 3rd — so ditch those two angles.

So you've got solutions $\frac{\pi}{3}$ and $\frac{5\pi}{3}$ in the range 0 to 2π. But you need all the solutions in the range -2π to 4π. Get these by repeatedly adding or subtracting 2π onto each until you go out of range:

$x = \frac{\pi}{3} \Rightarrow$ (adding 2π) $x = \frac{7\pi}{3}, \frac{13\pi}{3}$ (too big)
and (subtracting 2π) $x = -\frac{5\pi}{3}, -\frac{11\pi}{3}$ (too small)

$x = \frac{5\pi}{3} \Rightarrow$ (adding 2π) $x = \frac{11\pi}{3}, \frac{17\pi}{3}$ (too big)
and (subtracting 2π) $x = -\frac{\pi}{3}, -\frac{7\pi}{3}$ (too small)

So the solutions are: $x = \frac{-5\pi}{3}, \frac{-\pi}{3}, \frac{\pi}{3}, \frac{5\pi}{3}, \frac{7\pi}{3}$ and $\frac{11\pi}{3}$

Solving Trig Equations in a Given Interval

Equations involving *(ax + b)* take a bit more *Work*

For these, you (usually) need to get your first solution from a calculator. Then you have a choice how you get the others — you can either draw a graph of $y = \sin(ax + b)$, or you can use a CAST diagram.

Example: Solve the equation $\sin(2x + 0.4) = 0.75$, for $0 \le x \le 2\pi$, giving your answer to 3 decimal places.

You need to check angles in the range $(2 \times 0 + 0.4)$ to $(2 \times 2\pi + 0.4)$ — that is 0.4 to 12.966.
From your calculator (set in radian mode) the first solution is 0.8481 (4 decimal places).

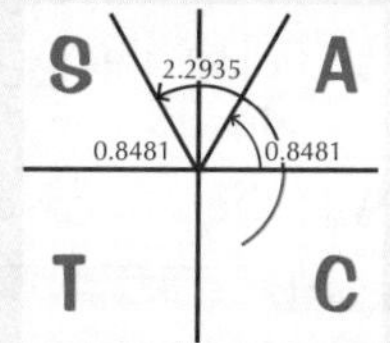

The angles which are positive for sin are in the 1st and 2nd quadrants.
The angles in the range 0 to 2π are **0.8481** and **2.2935** ($\pi - 0.8481$).

Now add and subtract 2π to each of these angles, chucking away any which are outside the range 0.4 to 12.966.

$2x + 0.4 = 0.8481 \Rightarrow$ (adding 2π) $2x + 0.4 =$ **7.1313**, 13.4144 (too big)
and (subtracting 2π) $2x + 0.4 = -5.4351$ (too small)
$2x + 0.4 = 2.2935 \Rightarrow$ (adding 2π) $2x + 0.4 =$ **8.5767**, 14.8599 (too big)
and (subtracting 2π) $2x + 0.4 = -3.9897$ (too small)

So $2x + 0.4 = 0.8481$, 2.2935, 7.1313 and 8.5767.
Now solve each equation to get the final answers: The solutions are $x = 0.224$, 0.947, 3.366 and 4.088.

You could get a *Decimal Answer* or a *Multiple of* π

Read the question carefully — they could ask you to round the answer to a certain number of decimal places, or maybe they'll make your life easier and tell you to leave it in terms of π.

Example: Solve $\cos^2 x = ½$ for $0 \le x \le 2\pi$

There are really two questions here, because when you take the square root there are two answers:

$$\cos x = \frac{1}{\sqrt{2}} \text{ and } \cos x = -\frac{1}{\sqrt{2}}$$

When you draw the CAST diagram the angles go in all 4 quadrants for positive and negative.

So the solutions from the CAST diagram are: $\frac{\pi}{4}, \frac{3\pi}{4}, \frac{5\pi}{4}$ and $\frac{7\pi}{4}$.

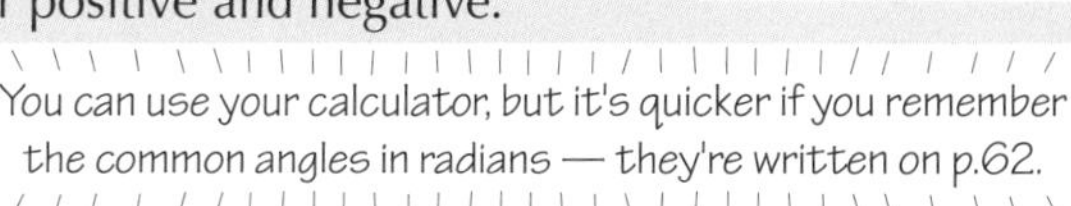

Adding or subtracting 2π to $\frac{\pi}{4}$ and $\frac{3\pi}{4}$ gives answers outside the required range, so these are the only answers.

Practice Questions

1) ***Solve each of these for $0 \le x \le 2\pi$***
   a) $\cos x = \frac{\sqrt{3}}{2}$ b) $\sin x = -\frac{1}{\sqrt{2}}$ c) $\tan x = -1$

2) ***Solve each of these for $-\pi \le \theta \le \pi$, giving your answers to 2 decimal places.***
   a) $\tan 3\theta = 1.2$ b) $\sin(\theta - 0.7) = ¼$ c) $\cos(2\theta + \frac{\pi}{4}) = 0.15$

3) ***Sample exam question:***

Solve $\sin(2\theta - \frac{\pi}{3}) = ½$ in the interval $0 \le \theta \le 2\pi$. [7 marks]

Oh the wonderful thing about trig, is trig is a wonderful thing...

This is definitely AS Level stuff. It can be pretty tough, since there's a lot to remember. You need to make sure you do everything in the correct order, find all the possible solutions, and deal with radians properly (including setting your calculator to work in radians/degrees as necessary). Hmmm, AS Maths — not one for the faint-hearted.

Inverse Trig Functions & Sec, Cosec and Cot

Skip these two pages if you're doing OCR A, OCR B, AQA A or AQA B.

More weird trig words to learn, but they're all just upside-down or back-to-front versions of old ones.

$\sin^{-1}$, $\cos^{-1}$ and $\tan^{-1}$ are the Inverses of sin, cos and tan

You just need to think backwards for these functions — e.g. $\sin^{-1} 0.5$ means 'the angle whose sine is 0.5'.

You can't draw the inverse of the full graphs of sin, cos and tan. You have to restrict the domains (the x-axis) so that every x-value has only one corresponding y-value (and vice versa). Then the functions are called one to one.

Start with the sin, cos and tan graphs you've seen a million times:

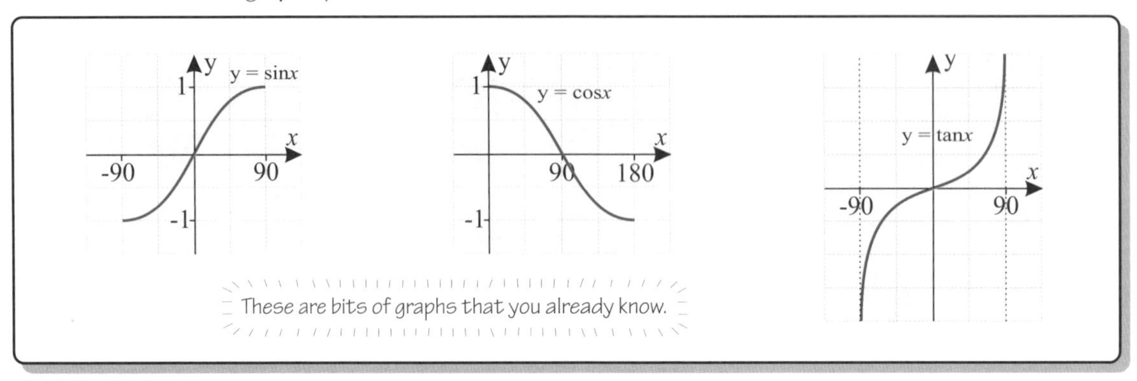

The graphs of the inverse functions are the reflections of the original graphs in the line y = x.

You need to learn these graphs off by heart.

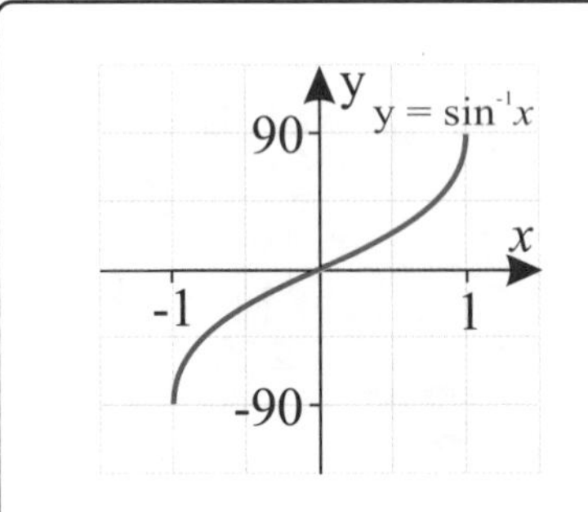

Domain is $-1 \le x \le 1$

Range is $-90° \le \sin^{-1}x \le 90°$

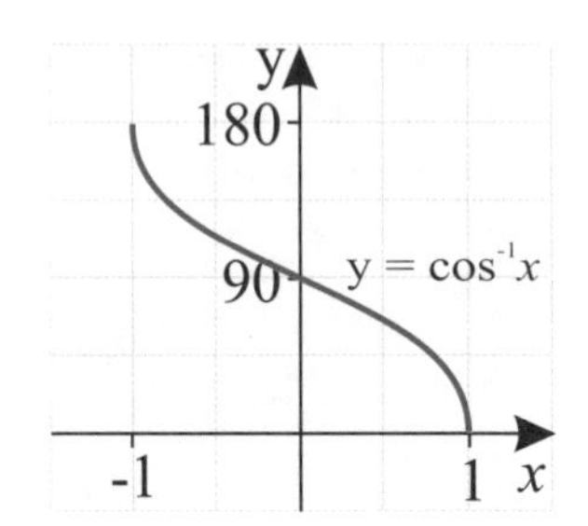

Domain is $-1 \le x \le 1$

Range is $0 \le \cos^{-1}x \le 180°$

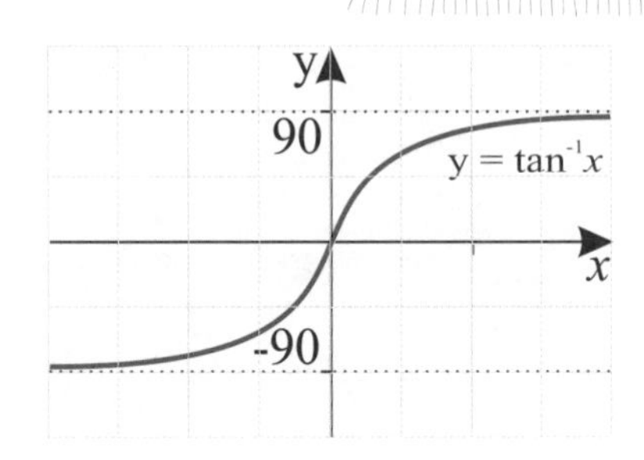

Domain is $x \in \mathbb{R}$

Range is $-90° \le \tan^{-1}x \le 90°$

Example:

Find the value of:

a) $\sin^{-1}0.35$ in degrees to 1 d.p.

b) $\cos^{-1}(-\frac{1}{\sqrt{2}})$ in radians in terms of π

a) This is like solving $\sin x = 0.35$, but you only need one solution — couldn't be easier:

$\sin^{-1}0.35 = 20.5°$ to 1 d.p.

Check your answer by working backwards — $\sin 20.5 = 0.35$

b) This is nastier, but there's no getting away from it — so just get stuck in.

Remember that $\cos\frac{\pi}{4} = \frac{1}{\sqrt{2}}$ — so that means that $\cos^{-1}(\frac{1}{\sqrt{2}}) = \frac{\pi}{4}$

Now, deal with the minus sign. The question asks for the value in the range 0 to π.

This gives $\cos^{-1}(-\frac{1}{\sqrt{2}}) = \frac{3\pi}{4}$

Inverse Trig Functions & Sec, Cosec and Cot

Sec, Cosec and Cot are Upside-down versions of Cos, Sin and Tan

sec x is short for secant x

$$\sec x = \frac{1}{\cos x}$$

cosec x is short for cosecant x

$$\text{cosec}\, x = \frac{1}{\sin x}$$

cot x is short for cotangent x

$$\cot x = \frac{1}{\tan x}$$

There's a handy third-letter rule to help you remember these functions: sec – cos, cot – tan, cosec – sin

Example: Find the values of a) sec 140° b) cosec $\frac{\pi}{5}$

a) $\sec 140 = \frac{1}{\cos 140°} = -1.31$

b) $\text{cosec}\frac{\pi}{5} = \frac{1}{\sin\frac{\pi}{5}} = 1.70$

You can work out the Graphs of sec, cosec and cot from cos, sin and tan

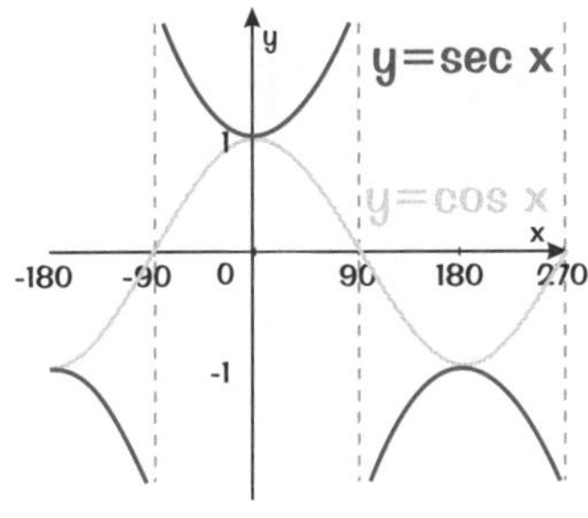

$-1 \le \cos x \le 1$ so $\sec x \ge 1$ or $\sec x \le -1$

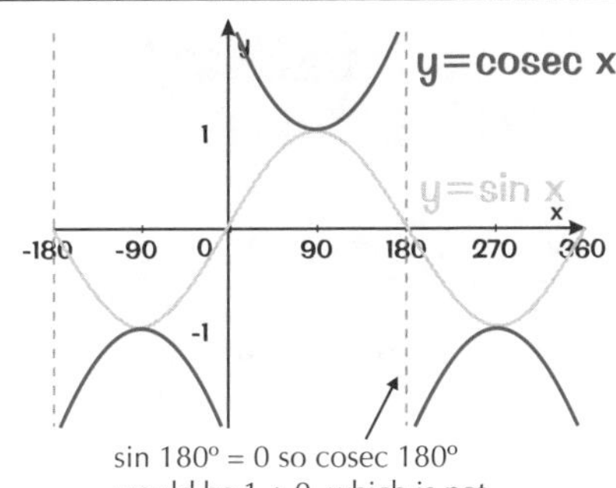

sin 180° = 0 so cosec 180° would be 1 ÷ 0, which is not allowed — so there's an asymptote

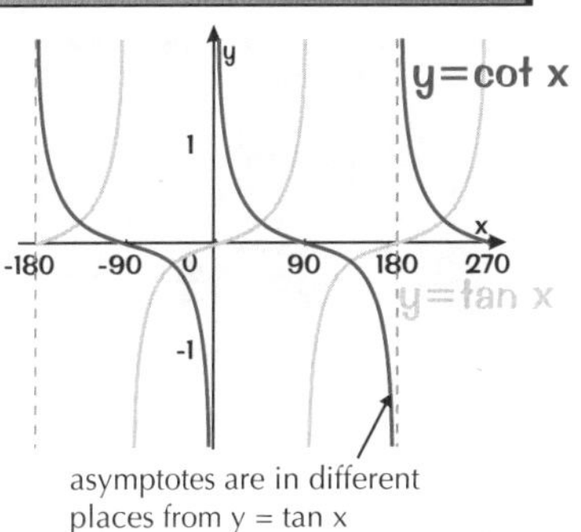

asymptotes are in different places from y = tan x

Example: Sketch the graph of $y = 2\sec(x + \frac{\pi}{2})$ for $-\pi \le x \le \pi$

This looks horrid, but take it slowly and do it in stages.

First sketch y = cos x (because $\sec x = \frac{1}{\cos x}$)

Next sketch y = sec x. Next sketch y = 2sec x.

Finally, shift the graph $\frac{\pi}{2}$ radians to the left.

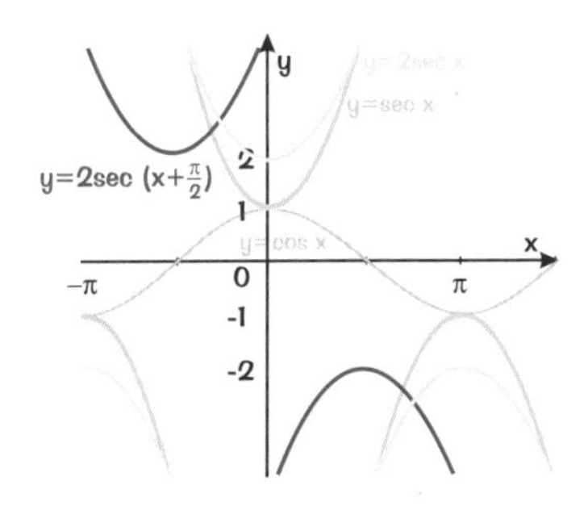

Practice Questions

1) ***Find the value in radians, in terms of π, of:*** ***a)*** $\tan^{-1}(-1)$ ***(for*** $-\frac{\pi}{2} \le \theta \le \frac{\pi}{2}$***)*** ***b)*** $\sin^{-1}\frac{\sqrt{3}}{2}$ ***(for*** $-\frac{\pi}{2} \le \theta \le \frac{\pi}{2}$***)***
2) ***Sketch the graph of*** $y = 90° - \cos^{-1} x$.
3) ***Solve for*** $0 \le x \le 360°$: ***a)*** $5\text{cosec}\, x + 8 = 0$ ***b)*** $\sec x - 2\cos x = 1$
4) ***Sample exam question:***

a) Sketch, on the same axes, graphs of i) y = sin 2x ii) y = cosec 2x – 1 for 0 ≤ x ≤ 360°. [3 marks]

b) Solve the equation sin 2x = cosec 2x – 1 for 0 ≤ x ≤ 360°, giving your answers correct to 1 decimal place. [7 marks]

If you only did AS Maths for the secs — you're going to be disappointed...

There's more about domains, ranges and inverse functions in Section One. And don't be fazed by 'sec', 'cosec', etc. They're just complicated ways of writing '1 ÷ cos' and so on. But remember which goes with which — cos starts with 'c', but its partner (sec) starts with 's'. Similarly, sin starts with 's', but it goes with cosec, which starts with 'c'. Confusing.

Identities with Sec, Cosec and Cot

Skip these two pages if you're doing OCR A, OCR B, AQA A or AQA B.

These give ways of writing one trig function in terms of others — the idea is to make problems easier. Sounds unlikely, but it actually works, and these ones aren't too hard to learn either.

The Identities $\tan x = \frac{\sin x}{\cos x}$ and $\cot x = \frac{\cos x}{\sin x}$ are Really Useful

You already know that $\tan x = \frac{\sin x}{\cos x}$. Well, since $\cot x = \frac{1}{\tan x}$ that means that $\cot x = \frac{\cos x}{\sin x}$.

Example: Solve the equation $\cot x = 2\cos x$ for $0 \le x \le 360°$

Start by putting cot in terms of cos and sin:

$$\frac{\cos x}{\sin x} = 2\cos x$$

($\cot x = \frac{\cos x}{\sin x}$)

$$\cos x = 2\cos x \times \sin x$$

Make one side equal to zero:

$$2\cos x \times \sin x - \cos x = 0$$

$$\cos x(2\sin x - 1) = 0$$

Then solve to get the possible x-values:

So $\cos x = 0$ or $2\sin x - 1 = 0$ ($\sin x = ½$)

$x = 90°$ or $270°$ $x = 30°$ or $150°$

Example: Prove the identity $\cot x + \tan x = \operatorname{cosec} x \sec x$

To prove an identity, always start with the left-hand side and change the way it looks, until it's the same as the right-hand side.

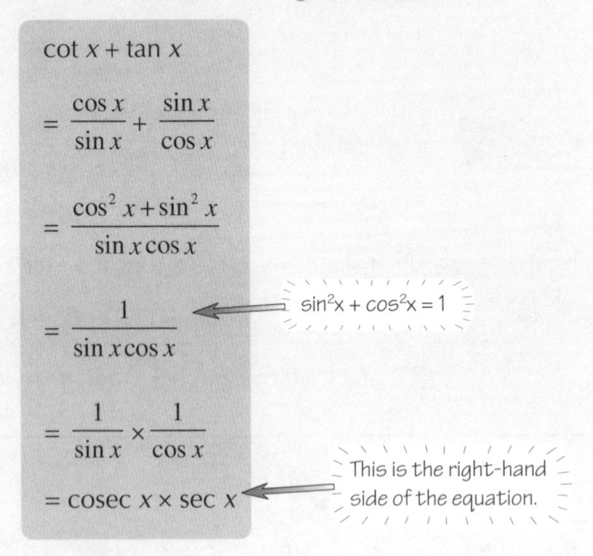

Start with the left-hand side: $\cot x + \tan x$

Replace cot and tan with sin and cos: $= \frac{\cos x}{\sin x} + \frac{\sin x}{\cos x}$

Put everything over a common denominator: $= \frac{\cos^2 x + \sin^2 x}{\sin x \cos x}$

Tidy everything up until it looks like the right-hand side: $= \frac{1}{\sin x \cos x}$ ($\sin^2 x + \cos^2 x = 1$)

$= \frac{1}{\sin x} \times \frac{1}{\cos x}$

$= \operatorname{cosec} x \times \sec x$ (This is the right-hand side of the equation.)

$\sin^2 x + \cos^2 x = 1$ gives you Two More Identities

Yep, you heard me — two spanking new identities from little old $\sin^2 x + \cos^2 x = 1$.

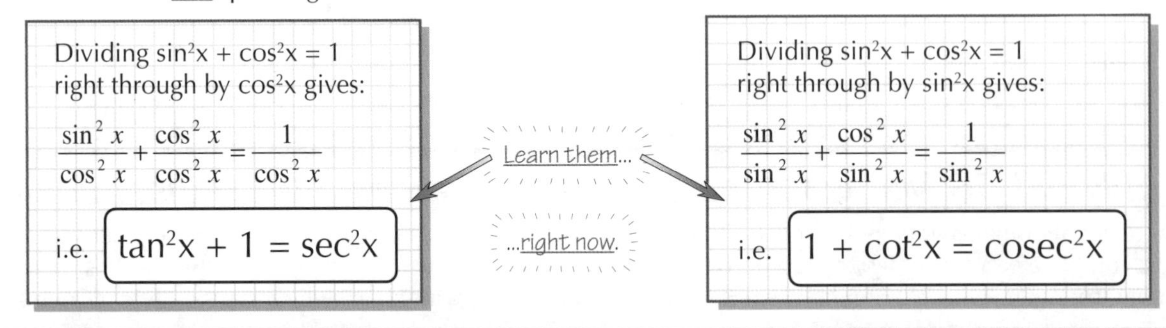

Dividing $\sin^2 x + \cos^2 x = 1$ right through by $\cos^2 x$ gives:

$$\frac{\sin^2 x}{\cos^2 x} + \frac{\cos^2 x}{\cos^2 x} = \frac{1}{\cos^2 x}$$

i.e. $\tan^2 x + 1 = \sec^2 x$

Dividing $\sin^2 x + \cos^2 x = 1$ right through by $\sin^2 x$ gives:

$$\frac{\sin^2 x}{\sin^2 x} + \frac{\cos^2 x}{\sin^2 x} = \frac{1}{\sin^2 x}$$

i.e. $1 + \cot^2 x = \operatorname{cosec}^2 x$

Learn them... ...right now.

Identities with Sec, Cosec and Cot

Use the **Identities** if you're given a **Squared Trig Function**

As soon as you see a squared trig function like $\sec^2 x$, you should be working out which identity to use to convert it.

Example: Solve, for $-180° \le x \le 180°$, these equations: a) $\text{cosec}^2 x + \cot^2 x = 2$ b) $\sec^2 x + \tan x = 3$

a) $\text{cosec}^2 x + \cot^2 x = 2$

$(1 + \cot^2 x) + \cot^2 x = 2$ ⟸ Use $1 + \cot^2 x = \text{cosec}^2 x$

$1 + 2\cot^2 x = 2$

$2\cot^2 x = 1$

$\cot^2 x = \frac{1}{2}$

$\tan^2 x = 2$ ⟸ $\tan x = \frac{1}{\cot x}$

$\tan x = \pm\sqrt{2}$

$x = \pm 54.7°$ or $\pm(180 - 54.7)$ so $x = \pm 54.7°$ or $\pm 125.3°$

b) $\sec^2 x + \tan x = 3$ ⟸ Use $\tan^2 x + 1 = \sec^2 x$

$\tan^2 x + 1 + \tan x = 3$

$\tan^2 x + \tan x - 2 = 0$ ⟸ This is a quadratic, and it factorises.

$(\tan x - 1)(\tan x + 2) = 0$

either: $\tan x = 1$, $x = 45°$ or $-135°$

or: $\tan x = -2$, $x = -63.4$ or $(180 - 63.4)$, $x = -63.4°$ or $116.6°$

You can also use them to prove other identities.

Example: Prove that $\cot^2 x + \cos^2 x = \text{cosec}^2 x - \sin^2 x$

left-hand side $= \cot^2 x + \cos^2 x$

$= \text{cosec}^2 x - 1 + \cos^2 x$ ⟸ Use $\cot^2 x = \text{cosec}^2 x - 1$

$= \text{cosec}^2 x - (1 - \cos^2 x)$

$= \text{cosec}^2 x - \sin^2 x$ ⟸ Use $1 - \cos^2 x = \sin^2 x$

Practice Questions

1) Solve the following equations for $0 \le x \le 360°$, giving your answers to 1 d.p.

a) $\cot x = 3\sin x$ b) $\text{cosec}^2 x = 2\cot x + 3$

2) Solve the following equations for $-\pi \le x \le \pi$, giving your answers in radians to 2 d.p.

a) $\cot^2 x = \text{cosec}\, x$ b) $\tan^2 x = 5 + \sec x$

3) Sample exam question:

a) Prove the identity $\tan^2 x - \sec x \tan x + 1 = \dfrac{1}{1 + \sin x}$ [6 marks]

b) Hence or otherwise solve the equation $\tan^2 x - \sec x \tan x = 1$, for $0 \le x \le 2\pi$, giving your solutions in radians in terms of π. [5 marks]

=):-)= This is text-speak for Abraham Lincoln... (Thought it might come in handy)

Okay, okay, I know — there is a lot to learn. The thing is (and this is the really important point here) — learning these will actually turn hard questions into easy ones. And doing easy questions is a far less stressful business than doing hard ones. So by buckling down now and learning these equations, you'll be doing yourself a favour in the long run.

Addition and Double Angle Formulas

Skip these two pages if you're doing OCR A, OCR B, AQA A or AQA B.

More formulas, but at least some of these are in the formula booklet. Make sure you know what's in the booklet and what isn't — then only learn by heart the stuff you really need to. That way you save brain power for where you really need it.

The **Addition Formulas** are for **sin(A ± B), cos(A ± B)** and **tan(A ± B)**

These beauties are in the formula booklet — make sure you know how to use them, but you don't need to learn them.

$$\sin(A + B) = \sin A \cos B + \cos A \sin B$$
$$\sin(A - B) = \sin A \cos B - \cos A \sin B$$

$$\cos(A + B) = \cos A \cos B - \sin A \sin B$$
$$\cos(A - B) = \cos A \cos B + \sin A \sin B$$

$$\tan(A + B) = \frac{\tan A + \tan B}{1 \quad \tan A \tan B}$$

$$\tan(A - B) = \frac{\tan A \quad \tan B}{1 + \tan A \tan B}$$

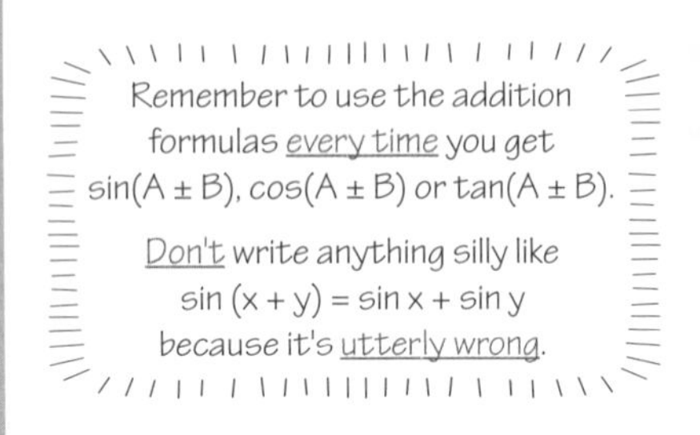

Example: Write, in surd form, the value of cos105°.

The trick is to get it in terms of more common angles. Why? Because you know their cos values in surd form.

105° = 60° + 45°, and you know cos of both of these in surd form — so start by converting cos105°.

See p.63 if you've forgotten.

$$\cos 105° = \cos(60° + 45°)$$
$$= \cos 60° \cos 45° - \sin 60° \sin 45°$$
$$= \frac{1}{2} \times \frac{1}{\sqrt{2}} - \frac{\sqrt{3}}{2} \times \frac{1}{\sqrt{2}}$$
$$= \frac{1-\sqrt{3}}{2\sqrt{2}} = \frac{\sqrt{2}-\sqrt{6}}{4}$$

Sketch the triangles on p.63 to help you remember these.

Example: Solve sin (x – 45°) = cosx for –180° ≤ x ≤ 180°

$$\sin(x - 45°) = \cos x$$ ← sin(A – B) = sinAcosB – cosAsinB

$$\sin x \cos 45° - \cos x \sin 45° = \cos x$$
$$\sin x \times \frac{1}{\sqrt{2}} - \cos x \times \frac{1}{\sqrt{2}} = \cos x$$
$$\sin x - \cos x = \sqrt{2}\cos x$$
$$\sin x = \sqrt{2}\cos x + \cos x$$
$$\sin x = (\sqrt{2} + 1)\cos x$$
$$\frac{\sin x}{\cos x} = \sqrt{2} + 1$$
$$\tan x = \sqrt{2} + 1$$
$$x = 67.5° \text{ or } -112.5°$$

You can **Work Out** the **Double-Angle Formulas** from the **Addition Formulas**

(...but it's best to learn them anyway)

Put B = A into the addition formulas and — hey presto — some *more* new formulas. Incredible.

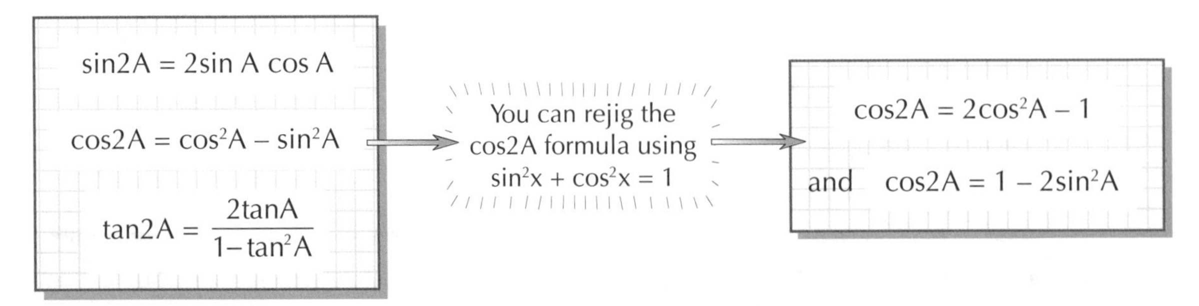

You need to know all five double angle formulas, or be able to work them out from the addition formulas. Use them when you need to turn an expression with a double angle into a single angle.

Addition and Double Angle Formulas

Example: Solve sin2x + sinx = 0 for $0 \le x \le 2\pi$, giving your answers in radians in terms of π.

$\sin 2x + \sin x = 0$

$2\sin x \cos x + \sin x = 0$ ← Use $\sin 2x = 2\sin x \cos x$

$\sin x(2\cos x + 1) = 0$

$\sin x = 0$ or $\cos x = -\frac{1}{2}$

$x = 0, \pi, 2\pi$ or $x = \frac{2\pi}{3}$ or $\frac{4\pi}{3}$

Example: Prove the identity cot 2x + cosec 2x = cot x

LHS = cot 2x + cosec 2x

$$= \frac{\cos 2x}{\sin 2x} + \frac{1}{\sin 2x} = \frac{\cos 2x + 1}{\sin 2x}$$

$$= \frac{(2\cos^2 x - 1) + 1}{\sin 2x} = \frac{2\cos^2 x}{2\sin x \cos x}$$

$$= \frac{\cos x}{\sin x} = \cot x = \text{RHS}$$

*The **Half–Angle Formulas** connect **sin½x** and **cos½x** with **cosx***

The half-angle formulas come from the identities for cos2A on p.70. You'll mainly use them when integrating.

Start with one of the cos 2A identities:

$\cos 2A = 1 - 2\sin^2 A$

Rearrange it: $2\sin^2 A = 1 - \cos 2A$

Get $\sin^2 A$ on its own: $\sin^2 A = \frac{1}{2}(1 - \cos 2A)$

Now let x = 2A: $\sin^2 \frac{1}{2}x = \frac{1}{2}(1 - \cos x)$

Do the same with the other cos 2A identity:

$\cos 2A = 2\cos^2 A - 1$

Rearrange it: $2\cos^2 A = 1 + \cos 2A$

Get $\cos^2 A$ on its own: $\cos^2 A = \frac{1}{2}(1 + \cos 2A)$

Now let x = 2A: $\cos^2 \frac{1}{2}x = \frac{1}{2}(1 + \cos x)$

Practice Questions

1) Find the exact value of tan 15°, giving your answer in the form $a + b\sqrt{3}$.

2) Solve $\sin x - \cos 2x = 1$ for $0 \le x \le 360°$, giving your answers in degrees to 1 d.p.

3) Solve $4\sin^2\frac{x}{2} + \cos^2 x = 3$ for $-\pi \le x \le \pi$, giving your answers in radians to 2 d.p.

4) Sample exam question:

a) Prove the identity $\frac{1 - \tan^2 x}{1 + \tan^2 x} = \cos 2x$. [4 marks]

b) Solve $\tan x \tan 2x = 1$ for $0 \le x \le 2\pi$, giving your answers in radians in terms of π. [5 marks]

If all else fails, multiply everything together and divide by π ...

These formulas are all about turning a question that you can't answer into one that you can. So if you find yourself bogged down trying to solve a trig equation that seems harder than anything you've practised on, think if you can replace anything. For example, you might be trying to solve something involving sinx cosx — this is hard, since there are two trig functions multiplied together. But if you replace this with ½sin2x — well, that's much nicer.

Addition and Double Angle Formulas

Skip these two pages if you're doing OCR A, OCR B, AQA A or AQA B.

Yep, another formula, and there's a method to learn this time as well. But it's not so bad once you get the hang of it.

You can write *a*cosθ + *b*sinθ as *R*sin(θ ± α) or *R*cos(θ ± α)

$$a\cos\theta + b\sin\theta = R\cos(\theta - \alpha)$$
$$a\cos\theta - b\sin\theta = R\cos(\theta + \alpha)$$
$$a\sin\theta + b\cos\theta = R\sin(\theta + \alpha)$$
$$a\sin\theta - b\cos\theta = R\sin(\theta - \alpha)$$

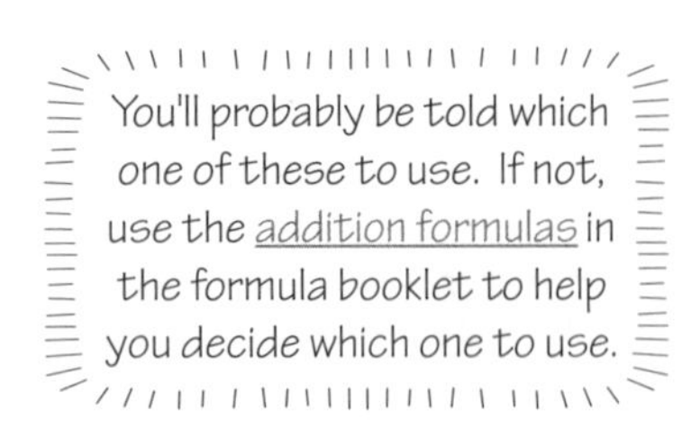

You can use any of the identities for any $a\cos\theta \pm b\sin\theta$, but stick to the ones in the box above and there'll be fewer negative values. Here's an example to show how to work out R and α.

$$a\cos\theta + b\sin\theta = R\cos(\theta - \alpha)$$
$$= R(\cos\theta\cos\alpha + \sin\theta\sin\alpha)$$ ← Using the addition formula (see p.70)
$$= R\cos\alpha\cos\theta + R\sin\alpha\sin\theta$$

You've got cosθ and sinθ on both sides, which means you can match up the bits in front of them:

$\mathbf{R\cos\alpha = a}$ and $\mathbf{R\sin\alpha = b}$

Work out α by dividing them: $\frac{R\sin\alpha}{R\cos\alpha} = \tan\alpha = \frac{b}{a}$ so $\alpha = \tan^{-1}\frac{b}{a}$

Now square them and add them together: $a^2 + b^2 = R^2(\cos^2\alpha + \sin^2\alpha) = R^2$ so $R = \sqrt{a^2 + b^2}$

Example: Write 8cos θ + 15sin θ in the form Rcos(θ – α) where R > 0 and $0 < \alpha < \frac{\pi}{2}$

Start with the form the question asks for:

$$8\cos\theta + 15\sin\theta = R\cos(\theta - \alpha)$$

Expand Rcos(θ – α) using the addition formula: $= R(\cos\theta\cos\alpha + \sin\theta\sin\alpha)$

$= R\cos\theta\cos\alpha + R\sin\theta\sin\alpha$

Compare with 8cos θ + 15sin θ: $\mathbf{R\cos\alpha = 8}$ and $\mathbf{R\sin\alpha = 15}$

Square the values and add them together: $R^2\cos^2\alpha + R^2\sin^2\alpha = 8^2 + 15^2$

$R^2(\cos^2\alpha + \sin^2\alpha) = 64 + 225$

Using $\cos^2\alpha + \sin^2\alpha = 1$ gives: $R^2 = 289$

$R = \sqrt{289} = 17$

Dividing gives $\frac{R\sin\alpha}{R\cos\alpha} = \tan\alpha = \frac{15}{8}$ so $\alpha = \tan^{-1}\left(\frac{15}{8}\right) = 1.0808$ radians

Putting it all together: $8\cos\theta + 15\sin\theta = 17\cos(\theta - 1.0808)$

Addition and Double Angle Formulas

*acosθ + bsinθ = Rcos(θ – α) can help you find **Max** and **Min** Values...*

Example: Find the maximum value of $8\cos\theta + 15\sin\theta$ and the smallest positive value of θ for which it occurs.

This is the same function as on the last page.

You already worked out that $8\cos\theta + 15\sin\theta = 17\cos(\theta - 1.0808)$

so the maximum value of $8\cos\theta + 15\sin\theta$ is the same as the maximum value of $17\cos(\theta - 1.0808)$

The max value of $17\cos(\theta - 1.0808)$ is 17, and occurs when $\cos(\theta - 1.0808) = 1$

$\theta - 1.0808 = 0$

$\theta = 1.0808$ radians

Because the maximum value of the cosine of any angle is 1.

*...and **Solve Equations** too*

Example: Solve the equation $3\sin\theta - 2\cos\theta = 1$ for $0 \leq \theta \leq 2\pi$, giving your solutions in radians to 2 d.p.

Write $3\sin\theta - 2\cos\theta = R\sin(\theta - \alpha) = R\sin\theta\cos\alpha - R\cos\theta\sin\alpha$

so $R\cos\alpha = 3$ and $R\sin\alpha = 2$

Squaring and adding gives $R^2 = 3^2 + 2^2$ so $R = \sqrt{13}$

Dividing gives $\tan\alpha = \frac{2}{3}$ so $\alpha = 0.588$

This gives $\mathbf{3\sin\theta - 2\cos\theta = \sqrt{13}\sin(\theta - 0.588)}$

So now you need to solve $\sqrt{13}\sin(\theta - 0.588) = 1$

$$\sin(\theta - 0.588) = \frac{1}{\sqrt{13}}$$

$\theta - 0.588 = 0.281$ or $\pi - 0.281$

$\theta = 0.281 + 0.588$ or $2.861 + 0.588$

$\theta = 0.87$ or 3.45 radians to 2 d.p.

Practice Questions

***1)** a) Write $5\cos\theta + 3\sin\theta$ in the form $R\cos(\theta - \alpha)$ where $R > 0$ and $0 < \alpha < 90°$.*

b) Find the maximum and minimum values of $5\cos\theta + 3\sin\theta$ and the smallest positive values of θ for which they occur.

***2)** By writing $6\sin\theta - 5\cos\theta$ in the form $R\sin(\theta - \alpha)$ solve the equation $6\sin\theta - 5\cos\theta = 4$ for $0 \leq \theta \leq 2\pi$.*

***3)** Sample exam question:*

$f(x) = 9\cos x - 3\sin x + 5$

a) Find the minimum value of $\frac{1}{f(x)}$ and the smallest positive value of x for which it occurs, giving your answers correct to 2 decimal places, and with x in degrees. [7 marks]

b) Solve the equation $f(x) = 3$ for $0 \leq x \leq 360°$ giving your solutions correct to the nearest degree. [4 marks]

The end is getting very close now, but don't be upset — there's always P3...

There, that wasn't too bad, was it? A nice little routine to follow. Like a lot of this stuff, the problem isn't always doing it — it's realising that that's what you have to do in the first place. Just remember that if you have sin's and cos's added together, you can always write this as either one sin or one cos — as long as there are no bits like sin2x or cos3x around.

More Identities

Skip these two pages if you're doing OCR A, OCR B, AQA A or AQA B.

These have got to be the nastiest identities out there. Some of them are given in the formula booklet, so at least you don't have to learn them all. But just using them is bad enough, I reckon. Anyway, take a deep breath and dive in.

The first set of **formulas** turn a **Product** into a **Sum**

These formulas all come from the addition formulas for $\sin(A \pm B)$ and $\cos(A \pm B)$:

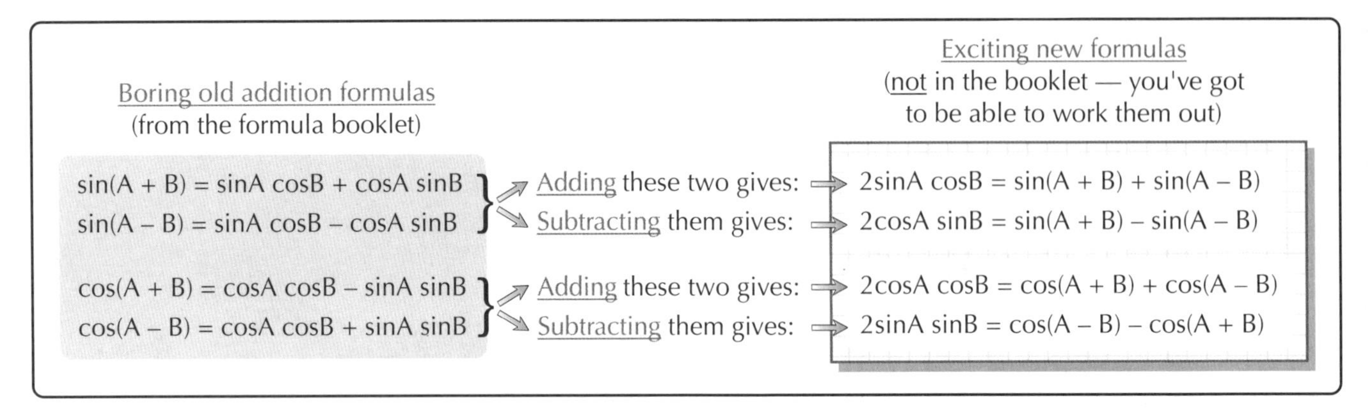

Example: Solve $4\cos 3x \cos x - 2\cos 2x = 1$ for $0 \le x \le \pi$

$$2(\cos 4x + \cos 2x) - 2\cos 2x = 1$$

$$2\cos 4x = 1$$

Using $2\cos A \cos B = \cos(A + B) + \cos(A - B)$ where $A = 3x$, $B = x$

$$\cos 4x = \frac{1}{2}$$

$$4x = \frac{\pi}{3}, \frac{5\pi}{3}, \frac{7\pi}{3}, \frac{11\pi}{3}$$

$$x = \frac{\pi}{12}, \frac{5\pi}{12}, \frac{7\pi}{12}, \frac{11\pi}{12}$$

The second set of formulas turn a **Sum** into a **Product**

These come from the formulas at the top of the page, putting $A + B = X$ and $A - B = Y$.

These ones are in the formula booklet — so you don't need to learn them, you lucky thing.

$$\sin X + \sin Y = 2\sin\frac{X+Y}{2}\cos\frac{X \quad Y}{2}$$

$$\sin X - \sin Y = 2\cos\frac{X+Y}{2}\sin\frac{X \quad Y}{2}$$

$$\cos X + \cos Y = 2\cos\frac{X+Y}{2}\cos\frac{X \quad Y}{2}$$

$$\cos X - \cos Y = -2\sin\frac{X+Y}{2}\sin\frac{X \quad Y}{2}$$

$A = \frac{(A+B)+(A \quad B)}{2} = \frac{X+Y}{2}$

$B = \frac{(A+B) \quad (A \quad B)}{2} = \frac{X \quad Y}{2}$

Watch out for the extra minus sign in the last one.

Example: Prove that $\sin 3x + \sin x = 4(\sin x - \sin^3 x)$

$$\text{LHS} = \sin 3x + \sin x = 2\sin\frac{3x+x}{2}\cos\frac{3x-x}{2}$$

Using $\sin X + \sin Y = 2\sin\frac{X+Y}{2}\cos\frac{X-Y}{2}$

$$= 2\sin 2x \cos x$$

$$= 2(2\sin x \cos x)\cos x$$

Using $\sin 2x = 2\sin x\cos x$

$$= 4\sin x \cos^2 x$$

Using $\sin^2 x + \cos^2 x = 1$

$$= 4\sin x(1 - \sin^2 x) = 4(\sin x - \sin^3 x) = \text{RHS}$$

More Identities

Example: Solve cos3x – cos5x = sinx for $0 \le x \le 180°$

cos3x – cos5x = sinx
-2sin4x sin(-x) = sinx
2sin4x sinx = sinx
2sin4x sinx – sinx = 0
sinx(2sin4x – 1) = 0
sinx = 0 or sin4x = ½
x = 0 or 180° or 4x = 30, 150, 390, or 510°
x = 0°, 7.5°, 37.5°, 97.5°, 127.5° or 180°

Take X = 3x and Y = 5x. Then $\cos X - \cos Y = -2\sin\frac{X+Y}{2}\sin\frac{X-Y}{2}$

Using sin (-x) = -sin x

Small angle approximations *only work if your angle is in* ***Radians***

When you've got a pretty small angle and it's measured in radians, these formulas are pretty good approximations.

$$\sin\theta \approx \theta \qquad \tan\theta \approx \theta \qquad \cos\theta \approx 1-\frac{\theta^2}{2}$$

If your angle's less than about 0.1 radians, these approximations are correct to 3 sig. figs.

Example: Using 3° ≈ 0.052 radians, find an approximation for: a) sin 63° b) cos 63° c) tan 63°

$$\begin{aligned}\sin 63° &= \sin(60°+3°)\\ &= \sin 60°\cos 3° + \cos 60°\sin 3°\\ &\approx \frac{\sqrt{3}}{2}\left(1-\frac{0.052^2}{2}\right)+\frac{1}{2}(0.052)\\ &\approx 0.891\end{aligned}$$

$$\begin{aligned}\cos 63° &= \cos(60°+3°)\\ &= \cos 60°\cos 3° - \sin 60°\sin 3°\\ &\approx \frac{1}{2}\left(1-\frac{0.052^2}{2}\right)-\frac{\sqrt{3}}{2}(0.052)\\ &\approx 0.454\end{aligned}$$

$$\tan 63° = \frac{\sin 63°}{\cos 63°} \approx \frac{0.891}{0.454} \approx 1.963$$

Practice Questions

1) Prove that 2cos3x sinx = sin2x(1 – 4sin²x)

2) Solve sin3x – sinx = 0 for $0 \le x \le 2\pi$

3) If θ is small enough that θ² can be ignored, show that $\tan\left(\frac{\pi}{4}+\theta\right) \approx \frac{1+\theta}{1-\theta}$

4) Sample exam question:

a) Prove that $\frac{\cos y - \cos x}{\sin x - \sin y} = \tan\frac{x+y}{2}$. [5 marks]

b) Solve $\frac{\cos x - \cos 3x}{\sin 3x - \sin x} + \tan x = 0$ for $0 \le x \le 2\pi$. [8 marks]

It is by studying that we attain as little misery and as much happiness as possible...

Aarrgghh... it's Dr Johnson again — the man was clearly a fool. But back to the maths... It's a good idea to have your formula booklet open whenever you have to use any of the trig formulas — especially the stinkers on these pages, as there are an awful lot of plus and minus signs flying around. Well then... that's it — the end of the book. The only other thing to say is that this P2 knowledge is very powerful indeed — use it wisely, and only for the good of mankind.

Answers

Section One — Functions

Page 3

1)

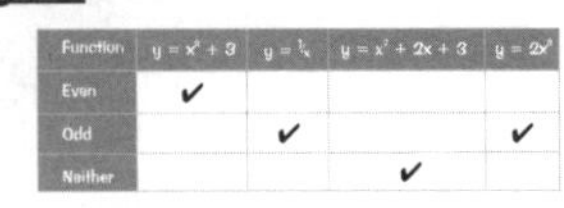

Function	$y = x^2 + 3$	$y = \frac{1}{x}$	$y = x^2 + 2x + 3$	$y = 2x^3$
Even	✔			
Odd		✔		✔
Neither			✔	

2) a) i) $f(x) \in \mathbb{R},\ f(x) \geq -2$ *[2 marks]*

ii) *An even function, since $f(-x) = f(x)$. [1 mark]*

b) *$g(x) = x^2 - 2 - 3x$. This is neither odd nor even since $f(-x)$ is not equal to $f(x)$ or $-f(x)$, e.g. $f(-1) = 2$, and $f(1) = -4$. [1 mark]*

Page 5

1) a) *-8*

b) *5*

c) *2*

2) a) *$hg(x) = 2(3x + 4)^2 = 18x^2 + 48x + 32$*

b) *$gh(x) = 3(2x^2) + 4 = 6x^2 + 4$*

c) *$hh(x) = 2(2x^2)^2 = 2(4x^4) = 8x^4$*

3) $h^{-1}(x) = +\sqrt{x+5}$ (the positive square root), where $x \in \mathbb{R}, x \geq -5$

4) a) i) *$f(-2) = (-2)^3 = -8$ [1 mark]*

ii) *$fg(-2) = (-2 - 2)^3 = (-4)^3 = -64$ [1 mark]*

iii) *$gf(-2) = (-2)^3 - 2 = -8 - 2 = -10$ [1 mark]*

b) i) *$fg(x) = (x - 2)^3 = x^3 - 6x^2 + 12x - 8$ [2 marks]*

ii) *$gf(x) = x^3 - 2$ [2 marks]*

c) $h^{-1}(x) = +\sqrt{4x+2}$ *[2 marks]* where $x \in \mathbb{R},\ x \geq -\frac{1}{2}$ *[1 mark]*

and $h^{-1}(x) \in \mathbb{R},\ h^{-1}(x) \geq 0$ *[1 mark]*

Page 7

1) a) $g(x) \in \mathbb{R},\ g(x) \geq -9$ *[2 marks]*

($g(x)$ has a u-shaped quadratic graph with zeros at $x = -2$ (although this is outside the domain of g) and $x = 4$, and so the minimum value is at $x = 1$ (halfway between the two zeros). This minimum value is -9, and $g(x)$ can take any value greater than or equal to this.)

b) *Complete the square:* $g(x) = (x-1)^2 - 9$

This means that $g^{-1}(x) = 1 + \sqrt{x+9}$ *[1 mark] where* $x \in \mathbb{R},\ x \geq -9$ *[1 mark]*

(this is the domain of $g^{-1}(x)$).

The range of $g^{-1}(x)$ is $g^{-1}(x) \in \mathbb{R},\ g^{-1}(x) \geq 1$ *[1 mark]*

(which is the same as the domain of $g(x)$).

c)

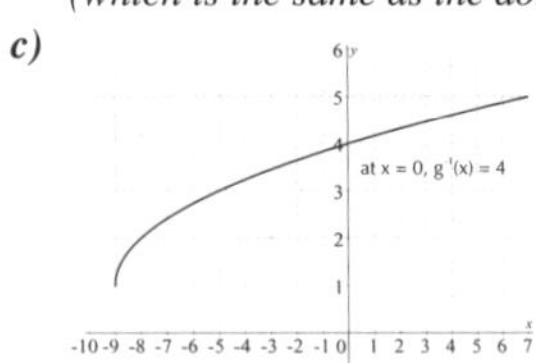

[2 marks for the shape, 1 mark for the correct intercept on the y-axis]

Page 9

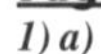

1) a)

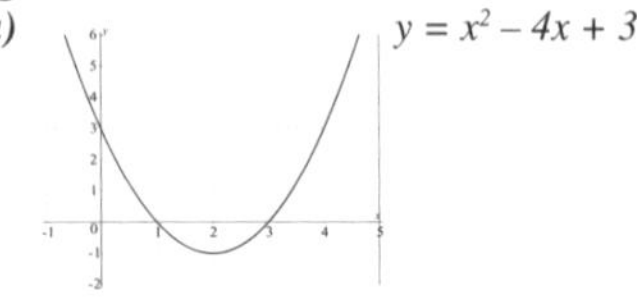

$y = x^2 - 4x + 3$

b)

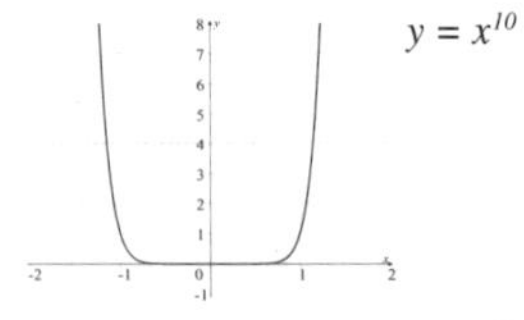

$y = x^{10}$

c)

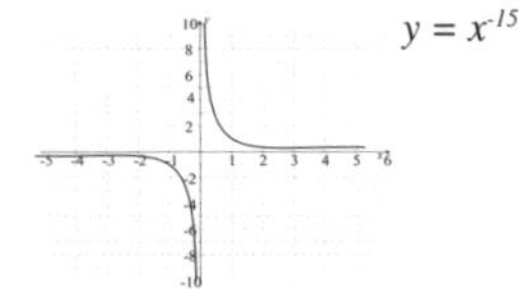

$y = x^{-15}$

d)

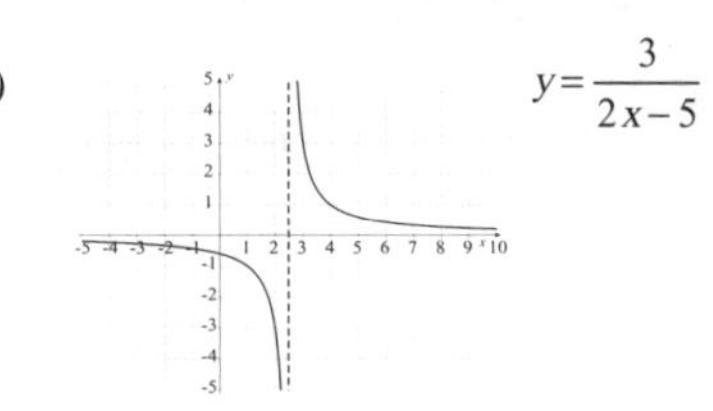

$y = \frac{3}{2x-5}$

2) *The asymptote is at $x = \frac{2}{3}$ — this is where the function is undefined (i.e. where the bottom line of the fraction is equal to zero) [1 mark].*

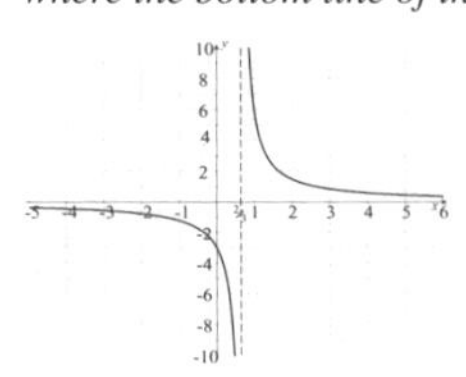

As $x \to \infty, y \to 0$ (and is positive). As $x \to -\infty, y \to 0$ (and is negative)

[3 marks for the graph — 1 mark for the correct general shape, plus 1 mark for the graph tending to each of the asymptotes correctly].

Page 11

1) a)

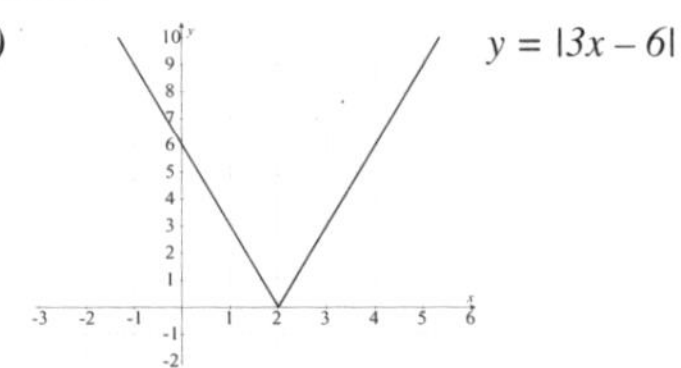

$y = |3x - 6|$

b)

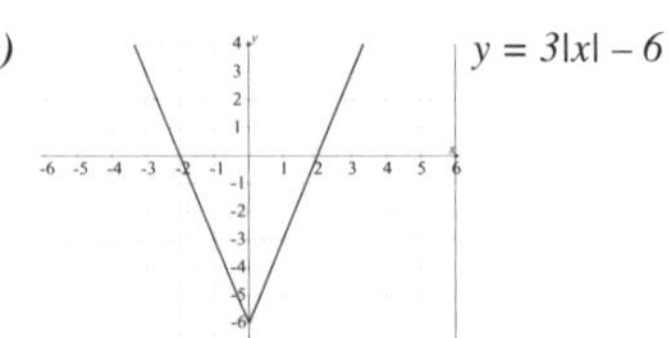

$y = 3|x| - 6$

2) a) $4|x| < 12 \qquad |x| < 3 \qquad$ so $-3 < x < 3$

b) $|x - 4| > 3$

$x - 4 < -3$ or $x - 4 > 3 \qquad$ so $\qquad x < 1$ or $x > 7$

c) *Draw a graph first:*

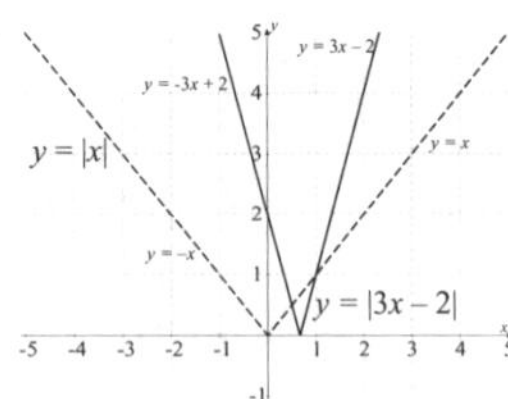

You need to find where the two graphs cross — you can see this is where the line $y = x$ meets the line $y = 3x - 2$ and where $y = x$ meets $y = -3x + 2$.

So first of all, solve $x = 3x - 2 \qquad$ i.e. $x = 1$

Then solve $x = -3x + 2 \qquad$ i.e. $x = \frac{1}{2}$.

So the lines cross at (1, 1) and $\left(\frac{1}{2}, \frac{1}{2}\right)$.

But you need the bits where the dotted line ($y = |x|$) is below the solid line ($y = |3x - 2|$), and so the answer is: $x > 1$ or $x < \frac{1}{2}$.

3) a) and ***b)***

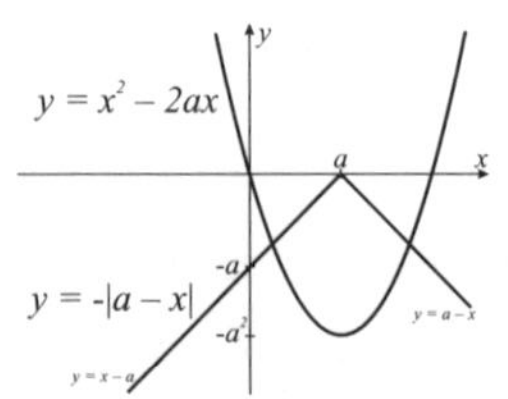

The easiest way to sketch the quadratic is to complete the square:

$y = x^2 - 2ax = (x-a)^2 - a^2$

So the parabola has its minimum at $x = a$, and the minimum value is $y = -a^2$ (which is less than $-a$, since a is greater than 1).

Answers

[2 marks for a correct answer to part a) — but 1 mark for getting half the graph correct]
[2 marks for a correct answer to part b) — but lose 1 mark if it's in the wrong position relative to the first graph]

c) Put $a = 2$. Then you need to find where $x^2 - 2ax = -|a - x|$, i.e. where $x^2 - 4x = -|2 - x|$ [1 mark].
The intersection point on the right is where the curve $y = x^2 - 4x$ meets the line $y = a - x$, (i.e. $y = 2 - x$).
So solve $x^2 - 4x = 2 - x$, or $x^2 - 3x - 2 = 0$ [1 mark]

This doesn't factorise, so use the quadratic formula: $x = \frac{-b \pm \sqrt{b^2-4ac}}{2a}$.

This gives $x = \frac{3 \pm \sqrt{17}}{2}$. But you know the answer you want is bigger than 2 (by looking at the graph and remembering that $a = 2$), so your first solution must be at $x = \frac{3+\sqrt{17}}{2}$ [1 mark].

You could find the other solution in a similar way (i.e. find where the line $y = x - a$ intersects the curve $y = x^2 - 4x$), but it's quicker if you notice that the two graphs are symmetrical about $x = 2$ [1 mark]. The first solution was a distance of $\frac{3+\sqrt{17}}{2} - 2 = \frac{\sqrt{17}-1}{2}$ to the right of $x = 2$, so the other solution must be the same distance to the left.

So the other solution is at $x = 2 - \frac{\sqrt{17}-1}{2} = \frac{5-\sqrt{17}}{2}$ [1 mark].

Page 13

1) a) $f(x) = (x + 2)(3x^2 - 10x + 15) - 36$
b) $f(x) = (x + 2)(x^2 - 3) + 10$
2) a) **(i)** You just need to find $f(-1)$. This is $-6 - 1 + 3 - 12 = -16$.
(ii) Now find $f(1)$. This is $6 - 1 - 3 - 12 = -10$.
b) **(i)** $f(-1) = -1$
(ii) $f(1) = 9$
3) If $f(x) = 2x^4 + 3x^3 + 5x^2 + cx + d$, then to make sure $f(x)$ is exactly divisible by $(x - 2)(x + 3)$, you have to make sure $f(2) = f(-3) = 0$.
$f(2) = 32 + 24 + 20 + 2c + d = 0$, i.e. $\underline{2c + d = -76}$.
$f(-3) = 162 - 81 + 45 - 3c + d = 0$, i.e. $\underline{3c - d = 126}$.
Add the two underlined equations to get: $5c = 50$, and so $\underline{c = 10}$.
Then $\underline{d = -96}$.

Page 15

1) a)

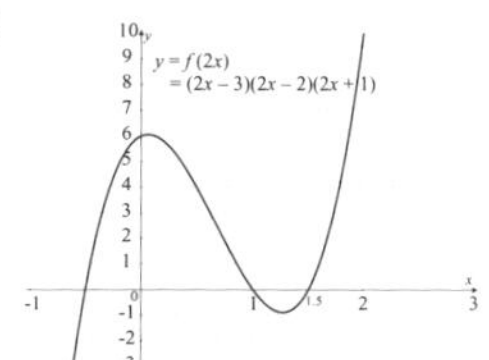

b)

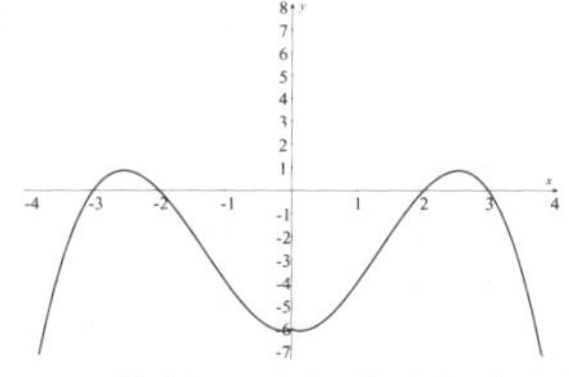

$y = -f(|x|) = -(|x| - 3)(|x| - 2)(|x| + 1)$

2) a) and **b)**

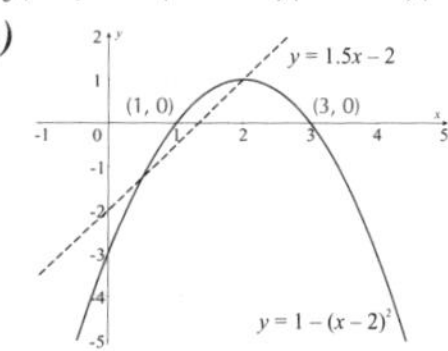

[3 marks in total for part a) — 1 mark for the correct basic shape, 1 mark for both points where the graph cuts the x-axis correct, and 1 mark for the correct maximum point]
[1 mark for part b)]

c) The lines meet where $1 - (x - 2)^2 = 1.5x - 2$ [1 mark]. This is where:

$$1-(x^2-4x+4)=1.5x-2$$
$$-x^2+4x-3=1.5x-2$$
$$x^2-2.5x+1=0, \text{ i.e. where } 2x^2-5x+2=0$$

[2 marks for correctly arriving at this last equation]

d) This quadratic factorises: $2x^2 - 5x + 2 = (2x-1)(x-2)$.
So the solutions are at $x = \frac{1}{2}$ and $x = 2$. [1 mark] Then put these values back into one of the original equations (use $y = 1.5x - 2$, since that's easier) to find that the points of intersection are $\left(\frac{1}{2}, -\frac{5}{4}\right)$ and $(2, 1)$
[1 mark for each correct point].

Section Two — Sequences and Series

Page 17

1) a) $u_1 = 1^2 + 3 = 4$; $u_2 = 2^2 + 3 = 7$; $u_3 = 3^2 + 3 = 12$; $u_4 = 4^2 + 3 = 19$
b) $u_{20} = 20^2 + 3 = 403$
2) $a_2 = 3 \times 4 - 2 = 10$; $a_3 = 3 \times 10 - 2 = 28$; $a_4 = 3 \times 28 - 2 = 82$
3) ($u_2 = 3 + 5 = 8$; $u_3 = 8 + 5 = 13$; $u_4 = 13 + 5 = 18$)
n^{th} term: $u_n = 5n - 2$

Page 19

1) $a = 5, d = 3, l = 65$
$a + (n - 1)d = l$
$5 + 3(n - 1) = 65$
$3(n - 1) = 60$
$n - 1 = 20$
$n = 21$

$S_{21} = 21\left(\frac{5+65}{2}\right) = 735$

2) a) $a + (n - 1)d = n$th term
$7 + (5 - 1)d = 23$
$4d = 16$
$d = 4$
b) $a + (15 - 1)d$ is 15th term
$= a + 14d = 7 + 14 \times 4 = 63$

c) $S_{10} = \frac{10}{2}[2 \times 7 + (10 - 1) \times 4]$

$S_{10} = 5(14 + 36) = 250$

3) $a + 6d = 36$
$a + 9d = 30$
Subtract one equation from the other:
$-3d = 6$
$\underline{d = -2}$
Plug d into one of the original equations:
$a + 6 \times -2 = 36$
$a - 12 = 36$
$\underline{a = 48}$

$S_5 = \frac{5}{2}[2 \times 48 + (5 - 1) \times -2]$

$S_5 = \frac{5}{2}(96 - 8)$ $\underline{=220}$

n^{th} term $= a + (n - 1)d$
$= 48 + (n - 1) \times -2$
$= 48 - 2n + 2 = \underline{50 - 2n}$

4) $\sum_{n=1}^{20}(3n-1) = 2 + 5 + 8 + \ldots + 59$

$= 20\left(\frac{2+59}{2}\right) = 610$

5) $\sum_{n=1}^{10}(48-5n)$
$= 43 + 38 + 33 + \ldots + -2$

$= 10\left(\frac{43+-2}{2}\right) = 205$

Answers

Page 20

1) a)

$l = a + (n-1)d$ [1 mark]. The average of the terms is $\frac{a+l}{2}$ [1mark].

The sum of the first n terms will be n multiplied by the average.

This is $S_n = n\frac{a+l}{2}$ [1 mark].

But $l = a + (n-1)d$, and so $S_n = \frac{n}{2}\left[2a + (n-1)d\right]$ [1 mark]

(You'll get the marks for any suitable method.)

b) $\sum_{n=9}^{32}(2n-5) = (2 \times 9 - 5) + (2 \times 10 - 5) + (2 \times 11 - 5) + ... + (2 \times 32 - 5)$

$= 13 + 15 + 17 + ... + 59$ [1 mark]

So $a = 13$, $l = 59$, $n = 32 - 8 = 24$ (32 terms minus the first 8) [1 mark]

$S_{24} = 24\left(\frac{13+59}{2}\right)$ (Using $S_n = n\frac{(a+l)}{2}$) [1 mark]

$S_{24} = 864$ [1 mark]

Page 23

1) $a = 2$, $r = -3$

10th term, $u_{10} = ar^9$

$= 2 \times (-3)^9 = -39366$

2) a) r = 2nd term ÷ 1st term

$r = 12 \div 24 = ½$

b) 7th term $= ar^6$

$= 24 \times (½)^6$

$= 0.375$ (or $\frac{3}{8}$)

c) $S_\infty = \frac{a}{1-r}$

$= \frac{24}{1-\frac{1}{2}}$

$= 48$

3) **G.P.** — 2, 6, ... $a=2$, $r=3$

5th term is $ar^4 = 2 \times 3^4 = \underline{162}$

A.P. — 2, 6, ... $a=2$, $d=4$

You need $a + (n-1)d = 162$

$2 + (n-1)4 = 162$

$4(n-1) = 160$

$n - 1 = 40$

$\underline{n = 41}$ i.e. the 41st term of the AP is equal to the 5th term of the G.P.

4) a) Second term is ar [1 mark]

Therefore $a \times \frac{-1}{2} = -2$ [1 mark]

So $a = 4$ [1 mark]

b) $4, -2, 1, \frac{-1}{2}, \frac{1}{4}, \frac{-1}{8}, \frac{1}{16}$

[1 mark for at least 3 terms correct, 2 marks for at least 5 correct, and 3 marks for all 7 correct]

c) $S_7 = \frac{4\left(1-\left(-\frac{1}{2}\right)^7\right)}{1-\frac{-1}{2}} = \frac{4\left(1+\frac{1}{128}\right)}{\frac{3}{2}} = \frac{2}{3} \times 4\left(\frac{129}{128}\right) = \frac{43}{16} = 2\frac{11}{16}$

[3 marks for the correct answer, otherwise up to 2 marks for some correct working]

d) $S_\infty = \frac{a}{1-r} = \frac{4}{1-\frac{-1}{2}} = \frac{8}{3} = 2\frac{2}{3}$

[3 marks for the correct answer, otherwise up to 2 marks for some correct working]

Page 25

1) a) $(1 + ax)^{10} = 1 + \frac{10}{1}(ax) + \frac{10\times9}{1\times2}(ax)^2 + \frac{10\times9\times8}{1\times2\times3}(ax)^3 + ...$ [1 mark]

$= 1 + 10ax + 45a^2x^2 + 120a^3x^3 + ...$ [1 mark]

b) $(2 + 3x)^5 = 2^5(1 + \frac{3}{2}x)^5$ [1 mark]

$= 2^5[1 + \frac{5}{1}\left(\frac{3}{2}x\right) + \frac{5\times4}{1\times2}\left(\frac{3}{2}x\right)^2 + ...]$

x^2 term is $2^5 \times \frac{5\times4}{1\times2}\left(\frac{3}{2}\right)^2 x^2$, so coefficient is $2^5 \times \frac{5\times4}{1\times2} \times \frac{3^2}{2^2} = 720$ [1 mark]

c) From a), coefficient of x^2 is $45a^2$

If it's equal to the x^2 coefficient in part b), then $45a^2 = 720$ [1 mark]

$a^2 = \frac{720}{45}$

$a^2 = 16$

$a = 4$ [1 mark]

(answer can't be -4, as part a) specified $a>0$)

Section Three — Powers, Logs and Exponentials

Page 27

1) a) m^{10}

b) r^2

c) s^{28}

2) a) $\sqrt{54} = \sqrt{9\times6} = 3\sqrt{6}$

b) $\sqrt{108} = \sqrt{9\times4\times3} = 3\times2\times\sqrt{3} = 6\sqrt{3}$

c) $\sqrt{\frac{5}{81}} = \frac{\sqrt{5}}{\sqrt{81}} = \frac{\sqrt{5}}{9}$

3) $\frac{1}{2}\sqrt{2} = \frac{\sqrt{2}}{\sqrt{2}\times\sqrt{2}} = \frac{1}{\sqrt{2}}$

4) a) $\frac{3}{1-\sqrt{2}} = \frac{3\times\left(1+\sqrt{2}\right)}{\left(1-\sqrt{2}\right)\left(1+\sqrt{2}\right)} = \frac{3+3\sqrt{2}}{1-2} = -3-3\sqrt{2}$

b) $\frac{\sqrt{5}+1}{\sqrt{5}-1} = \frac{\left(\sqrt{5}+1\right)\left(\sqrt{5}+1\right)}{\left(\sqrt{5}-1\right)\left(\sqrt{5}+1\right)} = \frac{5+2\sqrt{5}+1}{5-1} = \frac{6+2\sqrt{5}}{4} = \frac{3+\sqrt{5}}{2}$

Page 29

1) a) $3^3 = 27$ so $\log_3 27 = 3$

b) to get fractions you need negative powers

$3^{-3} = {}^1/_{27}$

$\log_3({}^1/_{27}) = -3$

c) logs are subtracted so divide

$\log_3 18 - \log_3 2 = \log_3(18 \div 2)$

$= \log_3 9$

$= 2$ $(3^2 = 9)$

2) a) logs are added so you multiply — remember $2 \log 5 = \log 5^2$

$\log 3 + 2 \log 5 = \log (3 \times 5^2)$

$= \log 75$

b) logs are subtracted so you divide and the power half means square root

$½ \log 36 - \log 3 = \log (36^{½} \div 3)$

$= \log (6 \div 3)$

$= \log 2$

3) This only looks tricky because of the algebra, just remember the laws

$\log_b(\chi^2 - 1) - \log_b(\chi - 1) = \log_b\{(\chi^2 - 1)/(\chi - 1)\}$

using the difference of two squares $(\chi^2 - 1) = (\chi - 1)(\chi + 1)$ and cancelling

$= \log_b(\chi + 1)$

4) a) Some marks are a give-away

$\log_3 3 = 1$ [1 mark]

b) You'll get the first mark for showing you can use one of the laws of logarithms and the other for successfully getting $\chi = 32$

$3 \log_a 2 = \log_a 2^3$

$\log_a 4 + 3 \log_a 2 = \log_a(4 \times 2^3)$

$= \log_a(4 \times 8)$

$= \log_a 32$

[2 marks for the correct answer, otherwise 1 mark for some correct working]

Answers

Page 31

1) *Start by taking the exponential of both sides:* $e^{\ln 7x} = e^3$
The exponential and log cancel out: $7x = e^3$
Divide both sides by 7 to get x on its own: $x = \frac{e^3}{7}$
$x = 2.869$

2) *Take the natural log of both sides:* $\ln e^{4x} = \ln 5$
The exponential and log cancel out: $4x = \ln 5$
Divide both sides by 4 to get x on its own: $x = \frac{\ln 5}{4}$
$x = 0.402$

3) *Combine this into 1 function using log rules:* $\ln 12x = 6$
Take the exponential of both sides: $e^{\ln 12x} = e^6$
The exponential and log cancel out: $12x = e^6$
Divide both sides by 12 to get x on its own: $\frac{e^6}{12}$
$x = 33.619$

4) *Combine this into 1 function using log rules:* $\ln 5x^2 = 7$
Take the exponential of both sides: $e^{\ln 5x^2} = e^7$
The exponential and log cancel out: $5x^2 = e^7$
Divide both sides by 5: $x^2 = \frac{e^7}{5}$
Take the square root of both sides to get x on its own: $x = \sqrt{\frac{e^7}{5}}$
$x = 14.810$ *(you only need the positive square root, since ln x isn't defined for* $x < 0$*).*

5) *Use exponential rules to combine into 1 function:* $e^{3x} = 6$
Take the natural log of both sides: $\ln e^{3x} = \ln 6$
The exponential and log cancel out: $3x = \ln 6$
Divide both sides by 3 to get x on its own: $x = \frac{\ln 6}{3}$
$x = 0.597$

6) $y = ab^x$, *and so* $\underline{\ln y = \ln a + x \ln b}$. *Plotting ln y against x gives you the following graph:*

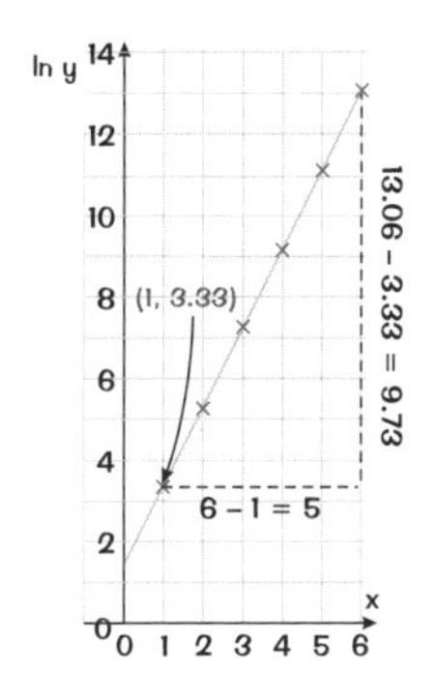

The gradient of the line is $9.73 \div 5 = 1.946$ *— this is ln b, and so* $b = e^{1.946} = 7$.
Then use the underlined equation and a point on the line (such as (1, 3.33)) to work out ln a:
$3.33 = \ln a + 1 \times 1.946$, *and so* $\ln a = 1.384$, *which means that* $a = e^{1.384} = 4$. *So* $y = 4 \times 7^x$.

Page 33

1) a) *Filling in the answers is just a case of using the calculator*

x	-3	-2	-1	0	1	2	3
y	0.0156	0.0625	0.25	1	4	16	64

Check that it agrees with what we know about the graphs. It goes through the common point (0,1), and it follows the standard shape.

b) *Then you just need to draw the graph, and use a scale that's just right.*

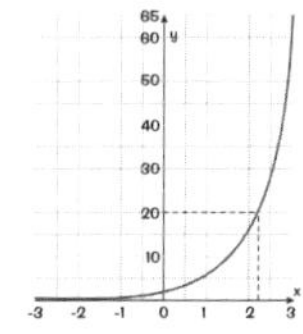

c) *The question tells you to use the graph to get your answer, so you'll need to include the construction lines, but check the answer with the calculator.*
$x = \ln 20 / \ln 4 = 2.16$, *but you can't justify this accuracy if your graph's not up to it, so 2.2 is a good estimate.*

2 a) $x = \log_{10}240 / \log_{10}10 = \log_{10}240 = 2.380$

b) $x = 10^{2.6} = 398.1$

c) $2x + 1 = \log_{10}1500 = 3.176$, *so* $2x = 2.176$, *so* $x = 1.088$

d) $(x - 1)\ln 4 = \ln 200$, *so* $x - 1 = \ln 200 / \ln 4 = 3.822$, *so* $x = 4.822$

3) *First solve for* $1.5^P = 1{,}000{,}000$
$P \times \log_{10}1.5 = \log_{10}1{,}000{,}000$,
so $P = (\log_{10}1{,}000{,}000) / (\log_{10}1.5) = 34.07$.
We need the next biggest integer, so this will be $P = 35$.

4) a) $2(5^{2x}) + 2(5^x) = 12$. *This looks tough, but we can try a couple of things.*
5^{2x} *is the same as* (5^x) *squared. If we call* $y = 5^x$, *then we get to the equation* $2y^2 + 2y = 12$.
We can rewrite this and solve as a quadratic equation.
$2y^2 + 2y - 12 = 0$. *[1 mark].*
Factorising the equation gives us $(2y + 6)(y - 2) = 0$, *so* $y = -3$ *or* $y = 2$.
So $5^x = -3$ *or 2. [1 mark]*
5^x *must be positive, so x will have no real value for* $5^x = -3$
So $5^x = 2$, *so* $x \log_{10}5 = \log_{10}2$,
so $x = \log_{10}2 / \log_{10}5 = 0.431$ *(3s.f.) [1 mark]*

b) *It may look difficult, but rewrite the equation using logs like the question says.*
$y = ab^x$ *gives us* $\log_{10}y = \log_{10}a + x\log_{10}b$. *This is so similar to* $y = mx + c$ *that it hurts.*
You need to think about the graph of $\log_{10}y$ *against x. The gradient will be* $\log_{10}b$, *the intercept will be* $\log_{10}a$ *[1 mark].*
Rewrite the table, but add another line for $\log_{10}y$

x	0	1	2	3	4
y	4	12	36	108	324
$\log_{10}y$	0.602	1.079	1.556	2.033	2.511

$\log_{10}a = 0.602$, *so* $a = 4$ *[1 mark]*
$\log_{10}b = \log_{10}12 - \log_{10}4$
$= \log_{10}(12/4)$
$= \log_{10}3$ *therefore* $b = 3$ *[1 mark]*
So the full equation is $y = 4 \times 3^x$

Section Four — Differentiation

Page 35

1 a) $\frac{dy}{dx} = nx^{n-1} = -4x^{-5}$

b) $\frac{dy}{dx} = nx^{n-1} = (3)(\frac{1}{2})x^{-\frac{1}{2}} = \frac{3}{2}x^{-\frac{1}{2}}$

c) *Apply the rule to each term one by one:*
$f'(x) = (4)(2)x^1 - (3)(-3)x^{-4}$
$f'(x) = 8x + 9x^{-4}$

2 a) i) *First write using index notation:*
$y = 3x^{\frac{1}{2}} + 4x^{-1}$
Now differentiate each term using the normal rule:
$$\frac{dy}{dx} = (3)(\tfrac{1}{2})x^{-\frac{1}{2}} + (4)(-1)x^{-2}$$
$$= \tfrac{3}{2}x^{-\frac{1}{2}} - 4x^{-2}$$

ii) *To get the gradient at (4,7) work out* $\frac{dy}{dx}$ *when* $x = 4$:
$$\frac{dy}{dx} = (\tfrac{3}{2})(4^{-\frac{1}{2}}) - (4)(4^{-2})$$
$$= (\tfrac{3}{2})(\tfrac{1}{2}) - (4)(\tfrac{1}{16})$$
$$= \tfrac{3}{4} - \tfrac{1}{4} = \tfrac{1}{2}$$

b) i) *Using index notation:* $y = 2x^{-2} + 3x^{-3}$
Differentiating each term one by one:
$$\frac{dy}{dx} = (2)(-2)x^{-3} + (3)(-3)x^{-4}$$
$$= -4x^{-3} - 9x^{-4}$$

ii) *To get the gradient at (1,5) work out* $\frac{dy}{dx}$ *when* $x = 1$:
$$\frac{dy}{dx} = (-4)(1^{-3}) - (9)(1^{-4})$$
$$= -4 - 9 = -13$$

3 a) *To verify that the graphs meet at (1,3) you only need to put* $x = 1$ *into each equation and show that the y values are both 3.*
For $y = 4 - x$, *at* $x = 1$:
$4 - x = 4 - 1 = \underline{3}$

For $y = \frac{2}{\sqrt{x}} + 1$, *at* $x = 1$:
$$\frac{2}{\sqrt{x}} + 1 = \frac{2}{1} + 1 = \underline{3}$$
Therefore the graphs must intersect at (1,3).

Answers

b) *To differentiate, write in index form:*
$y = 2x^{-\frac{1}{2}} + 1$
Now differentiate:
$\frac{dy}{dx} = (2)(-\frac{1}{2})x^{-\frac{3}{2}}$
$= -x^{-\frac{3}{2}}$

c) *To find the gradient of* $y = \frac{2}{\sqrt{x}} + 1$ *at (1,3), substitute x = 1 into the formula:* $\frac{dy}{dx} = -(1^{-\frac{3}{2}}) = -1$

Page 37

1) $y = e^x - \ln 4x$ $\quad \frac{dy}{dx} = e^x - \frac{1}{x}$
At a stationary point $e^x - \frac{1}{x} = 0$ so $e^x = \frac{1}{x}$
$xe^x = 1$ so $xe^x - 1 = 0$

2) $f(x) = 4e^x + 3 \ln 2x$
$f'(x) = 4e^x + \frac{3}{x}$ *[2 marks — 1 for each bit of the equation]*
$f'(1) = 4e^1 + \frac{3}{1}$
$= 4e + 3$ *[2 marks — 1 for each bit of the equation]*

Page 39

1) $y = e^{-2x} + 2x$
$\frac{dy}{dx} = -2e^{-2x} + 2$
At stationary point $0 = -2e^{-2x} + 2$
$-2 = -2e^{-2x}$
$1 = e^{-2x}$
$\therefore x = 0$ (as $e^0 = 1$)
When $x = 0$, $y = e^{-2x} + 2x = e^0 + 0 = 1$
Stationary point is (0, 1)

2) $\frac{d^2y}{dx^2} = 4e^{-2x}$
At $x = 0$, $\frac{d^2y}{dx^2} = 4e^{-2\times 0} = 4$, so stationary point is a minimum.

3) a) $y = x^2 - \ln 3x$
$\frac{dy}{dx} = 2x - \frac{1}{x}$ *[2 marks]*
At stationary point $0 = 2x - \frac{1}{x}$ *[1 mark]*
$2x = \frac{1}{x}$
$x^2 = \frac{1}{2}$ (question states $x > 0$, so only use positive square root)
$x = \frac{1}{\sqrt{2}}$ *[1 mark]*

b) $\frac{dy}{dx} = 2x - x^{-1}$
$\frac{d^2y}{dx^2} = 2 + x^{-2}$ *[1 mark]*
When $x = \frac{1}{\sqrt{2}}$, $\frac{d^2y}{dx^2} = 2 + (\frac{1}{\sqrt{2}})^{-2} = 4$ *[2 marks]*

c) P is a minimum as $\frac{d^2y}{dx^2} > 0$ *[1 mark]*

Page 41

1) *To find the gradient of the tangent, differentiate.*
Rewrite $\frac{2}{x}$ in index notation: $y = 2x^{-1}$
Differentiate: $\frac{dy}{dx} = -2x^{-2}$
At $x = 1$: $\frac{dy}{dx} = -2$
Use $y - y_1 = m(x - x_1)$ to write the equation of the tangent:
$y - 2 = -2(x - 1)$
$y = -2x + 4$

2) In index form $y = x^{\frac{1}{2}}$
$\frac{dy}{dx} = \frac{1}{2}x^{-\frac{1}{2}}$
At (4 , 2): $\frac{dy}{dx} = \frac{1}{4}$
So gradient of normal $= -1 \div \frac{1}{4} = -4$.
Now use $y - y_1 = m(x - x_1)$ to write the equation of the normal
$y - 2 = -4(x - 4)$
$y = -4x + 18$

3) a) At P, $x = 0$, so $y = e^0 - 4 \times 0 = 1$ So <u>P is (0, 1)</u> *[1 mark]*
At Q, $x = \ln 6$, so $y = e^{\ln 6} - 4 \ln 6 = 6 - 4 \ln 6$
<u>So Q is (ln 6, 6 – 4 ln 6)</u> *[1 mark]*

b) $y = e^x - 4x$
$\frac{dy}{dx} = e^x - 4$ *[1 mark]*
At P, $\frac{dy}{dx} = e^0 - 4 = -3$
Gradient of normal $= -1 \div -3 = \frac{1}{3}$ *[1 mark]*
Use $y = mx + c$:
So $y = \frac{1}{3}x + 1$
So equation of normal is $3y = x + 3$ *[2 marks]*

c) *Next you need to find the equation of the tangent at Q:*
$\frac{dy}{dx} = e^x - 4$
At Q, $\frac{dy}{dx} = e^{\ln 6} - 4 = 2$
Gradient of tangent is 2 *[1 mark]*
Use $y - y_1 = m(x - x_1)$ to write the equation of the tangent:
$y - (6 - 4 \ln 6) = 2(x - \ln 6)$
$y = 2x + 6 - 6 \ln 6$ *[1 mark]*
To find coordinates of R you must solve the simultaneous equations
$y = \frac{1}{3}x + 1$ and $y = 2x + 6 - 6 \ln 6$
$\frac{1}{3}x + 1 = 2x + 6 - 6 \ln 6$
$\frac{5}{3}x = 6 \ln 6 - 5$
$x = \frac{3}{5}(6 \ln 6 - 5)$ *[2 marks]*
To find the y-coordinate, substitute into the ***simpler*** *of the two equations:*
$y = \frac{1}{3}x + 1$
$y = \frac{1}{3}\{\frac{3}{5}(6 \ln 6 - 5)\} + 1 = \frac{6}{5}\ln 6$ *[2 marks]*

Page 43

1) a) $f(x) = uv$, where $u = x^3$ and $v = e^{2x}$.
So $u' = 3x^2$ and $v' = 2e^{2x}$.
So $f'(x) = 3x^2e^{2x} + x^3 \times 2e^{2x} = (3 + 2x)x^2e^{2x}$

b) $f(x) = uv$, where $u = x$ and $v = \ln x$.
So $u' = 1$ and $v' = 1/x$.
So $f'(x) = \ln x + x \times 1/x = \ln x + 1$

2) a) $f(x) = u/v$, where $u = x^2$ and $v = x + 1$.
So $u' = 2x$ and $v' = 1$.
So $f'(x) = \frac{2x(x+1) - x^2 \times 1}{(x+1)^2} = \frac{x^2 + 2x}{(x+1)^2} = \frac{x(x+2)}{(x+1)^2}$

b) $f(x) = u/v$, where $u = e^x$ and $v = \ln x$.
So $u' = e^x$ and $v' = 1/x$.
So $f'(x) = \frac{e^x \ln x - e^x \frac{1}{x}}{(\ln x)^2} = \frac{e^x}{(\ln x)^2}\left(\ln x - \frac{1}{x}\right)$

3) $y = \frac{e^x}{x^2} = \frac{u}{v}$, where $u = e^x$ and $v = x^2$.
This means $u' = e^x$ and $v' = 2x$, and so
$\frac{dy}{dx} = \frac{x^2e^x - 2xe^x}{x^4} = \frac{xe^x}{x^4}(x-2) = \frac{e^x}{x^3}(x-2)$ (since $x \neq 0$) [2 marks].
The graph has its turning points where $\frac{dy}{dx} = 0$ [1 mark].
But $x \neq 0$, and $e^x \neq 0$, and so the only turning point is at $x = 2$ [1 mark].
So the coordinates of the turning point are $\left(2, \frac{e^2}{4}\right)$ [1 mark]

Answers

Page 45

1 a) $y = (x + 2)^7$

$u = x + 2 \qquad \frac{du}{dx} = 1$

$y = u^7 \qquad \frac{dy}{du} = 7u^6$

$\frac{dy}{dx} = 7u^6 \quad = 7(x + 2)^6$

b) $y = (3x^2 + 5x)^4$

$u = 3x^2 + 5x \qquad \frac{du}{dx} = 6x + 5$

$y = u^4 \qquad \frac{dy}{du} = 4u^3$

$\frac{dy}{dx} = 4u^3(6x + 5) = 4(6x + 5)(3x^2 + 5x)^3$

c) $y = \sqrt[3]{2x-5}$

$u = 2x - 5 \qquad \frac{du}{dx} = 2$

$y = u^{1/3} \qquad \frac{dy}{du} = \frac{1}{3}u^{-2/3}$

$\frac{dy}{dx} = \frac{1}{3}u^{-2/3} \times 2 \quad = \frac{2}{3}(2x-5)^{-2/3}$

d) $y = \frac{1}{(4x+5)^3}$

$u = 4x + 5 \qquad \frac{du}{dx} = 4$

$y = u^{-3} \qquad \frac{dy}{du} = -3u^{-4}$

$\frac{dy}{dx} = -3u^{-4} \times 4 \quad = -12(4x + 5)^{-4}$

2) $y = \sqrt{(x + 5)} \quad = (x + 5)^{1/2}$

$u = x + 5 \qquad \frac{du}{dx} = 1$

$y = u^{1/2} \qquad \frac{dy}{du} = \frac{1}{2}u^{-1/2}$

$\frac{dy}{dx} = \frac{1}{2} u^{-1/2} \quad = \frac{1}{2} (x + 5)^{-1/2} = \frac{1}{2\sqrt{x+5}}$

When $x = -1$, $\frac{dy}{dx} = \frac{1}{4}$

so gradient of normal = $-1 \div \frac{1}{4} = -4$

When $x = -1$, $y = 2$ $(-1, 2)$

To find the equation of the normal, use $y - y_1 = m(x - x_1)$:

$y - 2 = -4(x - -1)$ and so $y = -4x - 2$

3 a) $y = \sqrt{(9 - 2x)}$

At A, $x = 0$, so $y = \sqrt{9} = 3$ So A is at (0,3) [1 mark]

At B, $y = 0$, so $9 - 2x = 0$, $x = 4.5$ So B is at (4.5,0) [1 mark]

b) *In index notation, $y = (9 - 2x)^{1/2}$*

Using chain rule to differentiate:

$u = 9 - 2x \qquad \frac{du}{dx} = -2$

$y = u^{1/2} \qquad \frac{dy}{du} = \frac{1}{2}u^{-1/2}$

$\frac{dy}{dx} = \frac{1}{2} u^{-1/2} \times -2$ *[2 marks]*

$\frac{dy}{dx} = -(9 - 2x)^{-1/2} \quad = \frac{-1}{\sqrt{9-2x}}$ *[1 mark]*

c) *At C, $x = 2.5$ and $y = 2$ So C is at (2.5,2)*

Gradient of tangent is $\frac{-1}{\sqrt{9-2\times2.5}} = -0.5$ [1 mark]

Equation of tangent at (2.5, 2):

$y - 2 = -0.5(x - 2.5)$

$y = -0.5x + 3.25$ [1 mark]

At D, $y = 0$, $-0.5x + 3.25 = 0$, so $x = 6.5$ So D is at (6.5, 0) [1 mark]

Page 47

1) *Area of circle:* $A = \pi r^2 \qquad \frac{dA}{dr} = 2\pi r$

Chain rule $\quad \frac{dA}{dt} = \frac{dA}{dr} \times \frac{dr}{dt}$

When $r = 2$ cm, $\frac{dA}{dr} = 2\pi r = 4\pi$ and the question tells you

$\frac{dA}{dt} = 3$ *cm²/hour*

$3 = 4\pi \times \frac{dr}{dt}$

$\frac{dr}{dt} = \frac{3}{4\pi} = 0.24$ *cm/hour*

2) *Volume of sphere* $V = \frac{4}{3}\pi r^3 \qquad \frac{dV}{dr} = 4\pi r^2$

Rate of change of volume: $\frac{dV}{dt} = 16$ *cm³ / s*

Chain rule: $\frac{dV}{dt} = \frac{dV}{dr} \times \frac{dr}{dt}$

$16 = 4\pi r^2 \times dr/dt$

When $V = 3000$, $r = \sqrt[3]{\frac{1}{\pi}\times\frac{3}{4}\times 3000}$ so $r = 8.95$ cm and $4\pi r^2 = 1005.9$cm²

$\frac{dr}{dt} = 16 \div 1005.9 \quad = 0.016$ *cm/s*

3) *Use the information about the volume to find the rate of change of the radius. Then examine the rate of change of the surface area.*

Volume of sphere: $V = \frac{4}{3}\pi r^3 \qquad \frac{dV}{dr} = 4\pi r^2$

Rate of change of volume: $\frac{dV}{dt} = 10$ *cm³ / s*

Chain rule: $\frac{dV}{dt} = \frac{dV}{dr} \times \frac{dr}{dt}$

$10 = 4\pi r^2 \times \frac{dr}{dt}$

When $r = 8$ cm, $4\pi r^2 = 804.25$

$\frac{dr}{dt} = 10 \div 804.25 \quad = 0.0124$ *cm/s*

The surface area of a sphere: $A = 4\pi r^2$ *so* $\frac{dA}{dr} = 8\pi r$

Rate of change of radius : $\frac{dr}{dt} = 0.0124$ *cm/s*

Chain rule: $\frac{dA}{dt} = \frac{dA}{dr} \times \frac{dr}{dt}$, *so* $\frac{dA}{dt} = 8\pi r \times \frac{dr}{dt}$

When $r = 8$ cm, $8\pi r = 201.06$ and $dr/dt = 0.0124$

$\frac{dA}{dt} = 201.06 \times 0.0124 \quad = 2.5$ *cm²/s*

Section Five — Integration

Page 49

1) a) $2x^5 + C$ **b)** $\frac{-2}{x^2} + C$ **c)** $\frac{3}{4}x^4 + \frac{2}{3}x^3 + C$

d) $3x^{4/3} + C = 3\sqrt[3]{x^4} + C$ **e)** $x^6 + \frac{2}{x} + \frac{2}{3}x^{3/2} + C$

2) $y = \frac{2}{3}x^{3/2} - \frac{2}{x} + C$

Putting $x = 1$ and $y = 0$ gives: $0 = \frac{2}{3} - 2 + C$

so $C = \frac{4}{3}$ *and the required curve is* $y = \frac{2}{3}x^{3/2} - \frac{2}{x} + \frac{4}{3}$

3) a) $\int_1^2 \frac{8}{x^5} + \frac{3}{\sqrt{x}}\,dx = \int_1^2 8x^{-5} + 3x^{-1/2}\,dx$

$= \left[\frac{8x^{-4}}{-4} + \frac{3x^{1/2}}{1/2}\right]_1^2 = \left[-\frac{2}{x^4} + 6\sqrt{x}\right]_1^2$

$= \left[(-\frac{1}{8} + 6\sqrt{2}) - (-2 + 6)\right] = -\frac{33}{8} + 6\sqrt{2}$

b) $\int_1^6 \frac{3}{y^2}\,dy = \int_1^6 3y^{-2}\,dy = \left[-\frac{3}{y}\right]_1^6$

$= \left[\left(-\frac{3}{6}\right) - \left(-\frac{3}{1}\right)\right] = \frac{5}{2}$

Answers

4) $\int_1^8 y\,dx = \int_1^8 x^{-\frac{1}{3}}\,dx = \left[\frac{3}{2}x^{\frac{2}{3}}\right]_1^8$

$= \left[\left(\frac{3}{2}\times 8^{\frac{2}{3}}\right)-\left(\frac{3}{2}\times 1^{\frac{2}{3}}\right)\right] = \left(\frac{3}{2}\times 4\right)-\left(\frac{3}{2}\times 1\right) = \frac{9}{2}$

Page 51

1) $\frac{2}{7}x^{\frac{7}{2}} + \frac{5}{2}e^x + C$

2) $\int\left(x+\frac{1}{x^2}\right)\left(x+\frac{1}{x^2}\right)dx = \int\left(x^2+\frac{2}{x}+\frac{1}{x^4}\right)dx$

$= \frac{x^3}{3} + 2\ln|x| - \frac{1}{3x^3} + C$

3) $\frac{3}{4}x^{\frac{4}{3}} - \frac{3}{2}e^x + \frac{4}{3}\ln|x| + C$ (or $\frac{3}{4}\sqrt[3]{x^4} - \frac{3}{2}e^x + \frac{4}{3}\ln|x| + C$)

4) $\int_1^3\left(4x-\frac{2}{x}\right)dx = \left[2x^2 - 2\ln|x|\right]_1^3$ *[3 marks]*

$= \left[(18 - 2\ln 3) - (2 - 2\ln 1)\right]$ *[2 marks]*

$= 16 - 2\ln 3$ *[2 marks]*

Page 53

1) $Volume = \pi\int y^2 dx = \pi\int_1^3 (x^2+1)^2 dx$

$= \pi\int_1^3 (x^4 + 2x^2 + 1)dx$ *[1 mark]*

$= \pi\left[\frac{x^5}{5} + \frac{2x^3}{3} + x\right]_1^3$ *[2 marks]*

$= \pi\left[\left(\frac{243}{5} + 18 + 3\right) - \left(\frac{1}{5} + \frac{2}{3} + 1\right)\right]$ *[2 marks]*

$= \frac{1016\pi}{15}$ *[1 mark]*

2) $y = \frac{4}{x^2}$ *but you need the equation in terms of x:*

Make x the subject so $x^2 = \frac{4}{y}$ *[2 marks]*

Volume $= \pi\int_1^2 x^2 dy = \pi\int_1^2 \frac{4}{y}dy$

$= \pi\left[4\ln|y|\right]_1^2$ *[2 marks]*

$= \pi\left[4\ln 2\right]$

$= 4\pi\ln 2$ *[2 marks]*

Page 55

1) a) $\frac{1}{28}(7x+2)^4 + C$ b) $-\frac{1}{4x-6} + C$ c) $\frac{1}{2}\ln|2x-3| + C$

d) $\frac{2}{5}e^{5x-1} + C$ e) $\frac{3}{4}\ln|4x+1| + C$ f) $-\frac{1}{(3x-2)^3} + C$

2) $\int_2^3 \frac{5}{2x-1}dx = \left[\frac{5}{2}\ln|2x-1|\right]_2^3$

$= \frac{5}{2}\ln 5 - \frac{5}{2}\ln 3 \qquad = \frac{5}{2}\ln\left(\frac{5}{3}\right)$

3) $\frac{dy}{dx} = \frac{2}{(4x-3)}$

Integrating gives $y = \frac{1}{2}\ln|4x-3| + C$ *[2 marks]*

To find the C, put in x = 1 and y = 3 to get $3 = \frac{1}{2}\ln 1 + C$ *[1 mark]*

So C = 3. [1 mark]

So the equation of the curve is $y = \frac{1}{2}\ln|4x-3| + 3$ *[1 mark]*

4) $\int y^2 dx = \int\left(1+\frac{1}{\sqrt{2x+1}}\right)^2 dx$

$= \int\left(1+\frac{1}{\sqrt{2x+1}}\right)\left(1+\frac{1}{\sqrt{2x+1}}\right)dx$

$= \int\left(1+\frac{2}{\sqrt{2x+1}}+\frac{1}{2x+1}\right)dx$ [2 marks]

$= \int\left(1+2(2x+1)^{-\frac{1}{2}}+\frac{1}{2x+1}\right)dx$ [1 mark]

$= x + \frac{2(2x+1)^{\frac{1}{2}}}{2\times\frac{1}{2}} + \frac{1}{2}\ln|2x+1| + C$

$= x + 2\sqrt{2x+1} + \frac{1}{2}\ln|2x+1| + C$ [2 marks]

Section Six — Numerical Methods

Page 57

1) $f(2) = 2^3 - 5\times 2 - 4 = 8 - 10 - 4 = -6 \; (<0)$

$f(3) = 3^3 - 5\times 3 - 4 = 27 - 15 - 4 = 8 \; (>0)$

Change of sign, so the root must lie in [2 , 3].

2) $f(4.315) = 4.315\ln 4.315 - 2 - 4.315 = -0.00605 \; (<0)$

$f(4.325) = 4.325\ln 4.325 - 2 - 4.325 = +0.00858 \; (>0)$

Change of sign, so the root is accurate to 2 decimal places.

3) a)

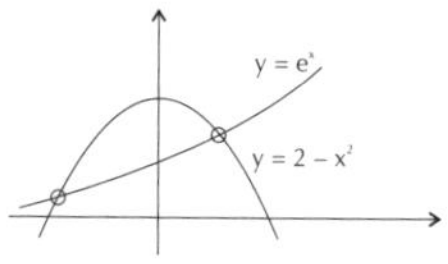

[1 mark for each correct line]

b) $2 - x^2 = e^x$ *and so* $2 - x^2 - e^x = 0$

$f(0) = 2 - 0 - e^0 = 1 \; (>0)$

$f(1) = 2 - 1 - e^1 = -1.718 \; (<0)$

Change of sign, so the root lies in the interval [0,1]

[2 marks for correct answer, otherwise 1 mark for some correct working]

c) *Try some values of n:*

If n= -2 then $f(-2) = 2 - 4 - e^{-2} = -2.13... \; (<0)$

If n= -1 then $f(-1) = 2 - 1 - e^{-1} = 0.63... \; (>0)$

Change of sign so root is in [-2, -1] — so if it's in [n, n+1] then n = -2.

[2 marks for correct answer, otherwise 1 mark for some correct working]

Page 59

1) $x^3 - 2x - 3 = 0$

so get the x^3 onto one side: $x^3 = 2x + 3$

and then take the cube root of both sides: $x = \sqrt[3]{(2x+3)}$

$x_0 = 2, x_1 = \sqrt[3]{(2\times 2+3)} = 1.912931183$

$x_2 = \sqrt[3]{(2\times 1.912931183+3)} = 1.896935259$

$x_3 = 1.893967062, x_4 = 1.89341526...$

The root = 1.893 to 3 d.p.

2) $x^2 - 5x + 2 = 0$

$f(0) = 0 - 0 + 2 = 2 \; (>0)$

$f(1) = 1 - 5 + 2 = -2 \; (<0)$

Change of sign, so the root lies in the interval (0 , 1).

$x^2 - 5x + 2 = 0$ *so* $5x = x^2 + 2$ *and* $x = \frac{(x^2+2)}{5}$.

(This makes p=2 and q=5.)

Starting with $x_0 = 0.5$:

$x_1 = \frac{(0.5^2+2)}{5} = 0.45$

$x_2 = \frac{(0.45^2+2)}{5} = 0.4405$

$x_3 = 0.43880805, x_4 = 0.4385105009...$

The root is 0.438 to 3 decimal places.

Answers

3) $f(x) = x^4 - 4x^3 + 2x^2 - 14x + 10$.
So $f'(x) = 4x^3 - 12x^2 + 4x - 14$.
Use $x_0 = 1$ as the first approximation, and use the Newton-Raphson formula to get a better one, x_1: $x_1 = 1 - \frac{f(1)}{f'(1)} = 1 - \frac{-5}{-18} = 0.722222$

Then use x_1 to get an even better approximation, x_2:

$$x_2 = 0.722222 - \frac{f(0.722222)}{f'(0.722222)} = 0.722222 - \frac{-0.302685}{-15.863510} = 0.703141$$

So the new approximation is $x_2 = 0.703$ (to 3 decimal places).

Page 61

1) a) $x_0 = 0$: $y_0 = \sqrt{9} = 3$
$x_1 = 1$: $y_1 = \sqrt{8} = 2.8284$
$x_2 = 2$: $y_2 = \sqrt{5} = 2.2361$
$x_3 = 3$: $y_3 = \sqrt{0} = 0$

$$h = \frac{(3-0)}{3} = 1$$

$$\int_a^b y\,dx = \frac{1}{2}[(3+0) + 2(2.8284 + 2.2361)] = 6.5645 \approx 6.56$$

b) $x_0 = -1$: $y_0 = e^1 = 2.7183$
$x_1 = -0.5$: $y_1 = e^{0.25} = 1.2840$
$x_2 = 0$: $y_2 = e^0 = 1$
$x_3 = 0.5$: $y_3 = e^{0.25} = 1.2840$
$x_4 = 1$: $y_4 = e^1 = 2.7183$
$x_5 = 1.5$: $y_5 = e^{2.25} = 9.4877$

$$h = \frac{(1.5+1)}{5} = 0.5$$

$$\int_a^b y\,dx \approx \frac{0.5}{2}[(2.7183 + 9.4877) + 2(1.2840 + 1 + 1.2840 + 2.7183)]$$

$$= 0.25 \times 24.7786 \approx 6.195$$

2) a) $p = \ln 2^1 = 0.6931$, $q = \ln 2^2 = 1.3863$ [1 mark for each]

b) $I \approx \frac{0.5}{2}[(0 + 1.3863) + 2(0.3466 + 0.6931 + 1.0397)]$
$= 1.386$ to 3 d.p.
[4 marks for the correct answer — otherwise up to 2 marks for substituting the numbers into the trapezium rule correctly, and 1 mark for doing the calculation correctly]

Section Seven — Trigonometry

Page 63

1) $L = r\theta = 20 \times \frac{\pi}{4} = 5\pi$ cm

$A = \frac{1}{2}r^2\theta = \frac{1}{2} \times 400 \times \frac{\pi}{4} = 50\pi$ cm^2

2) $r\theta = 7.56$ so $8\theta = 7.56 \Rightarrow \theta = 0.945$ radians
$A = \frac{1}{2}r^2\theta = \frac{1}{2} \times 64 \times 0.945 = 30.24$ cm^2

3) The area is made up of 2 segments:

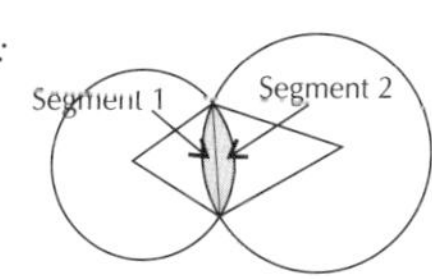

Area of segment 1 = $\frac{1}{2} \times 6^2 \times (\frac{\pi}{3} - \sin\frac{\pi}{3})$ [1 mark]

$= 18 \times (\frac{\pi}{3} - \frac{\sqrt{3}}{2})$ [1 mark]
$= 6\pi - 9\sqrt{3}$ [1 mark]

Area of segment 2 = $\frac{1}{2} \times 12^2 \times (\frac{\pi}{6} - \sin\frac{\pi}{6})$ [1 mark]

$= 72 \times (\frac{\pi}{6} - \frac{1}{2})$
$= 12\pi - 36$ [1 mark]
So the total area = $(18\pi - 36 - 9\sqrt{3})$ cm^2 [1 mark]

Page 65

1) **a)** $\frac{\pi}{6}, \frac{11\pi}{6}$

b) $\frac{5\pi}{4}, \frac{7\pi}{4}$

c) $\frac{3\pi}{4}, \frac{7\pi}{4}$

2) **a)** -2.85, -1.80, -0.76, 0.29, 1.34, 2.39
b) –2.69, 0.95
c) -2.82, -1.10, 0.32, 2.04

3) $2\theta - \frac{\pi}{3} = \frac{\pi}{6}, \frac{5\pi}{6}, \frac{13\pi}{6}, \frac{17\pi}{6}$ [3 marks]
(You should know that $\sin\frac{\pi}{6} = \frac{1}{2}$ (or you can use your calculator and convert the answer to a multiple of π). Then from the symmetry of a sine graph you can work out that another solution is $\frac{5\pi}{6}$. Then you have to add/subtract multiples of 2π to these answers until you get an answer for $2\theta - \frac{\pi}{3}$ that's out of the range you're interested in. The range for $2\theta - \frac{\pi}{3}$ that you're interested in is $-\frac{\pi}{3} \le 2\theta - \frac{\pi}{3} \le 4\pi - \frac{\pi}{3}$, which you get by multiplying the inequality $0 \le \theta \le 2\pi$ by 2, and subtracting $\frac{\pi}{3}$.)

$$2\theta = \frac{\pi}{6} + \frac{\pi}{3}, \frac{5\pi}{6} + \frac{\pi}{3}, \frac{13\pi}{6} + \frac{\pi}{3}, \frac{17\pi}{6} + \frac{\pi}{3}$$

i.e. $2\theta = \frac{\pi}{2}, \frac{7\pi}{6}, \frac{5\pi}{2}, \frac{19\pi}{6}$

$\theta = \frac{\pi}{4}, \frac{7\pi}{12}, \frac{5\pi}{4}, \frac{19\pi}{12}$ [4 marks]

Page 67

1) a) $\tan^{-1}(-1) = -\frac{\pi}{4}$

b) $\sin^{-1}(\frac{\sqrt{3}}{2}) = \frac{\pi}{3}$

2) First sketch $y = \cos^{-1}x$, then reflect it in x-axis to get $y = -\cos^{-1}x$
Finally shift it up 90 to give $y = -\cos^{-1}x + 90$ or $y = 90 - \cos^{-1}x$

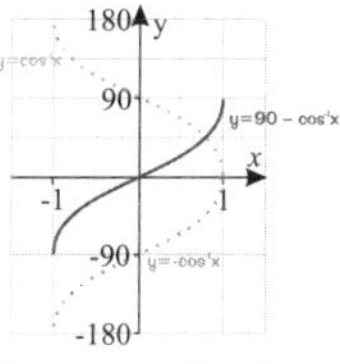

3) a) $5\text{cosec}\,x + 8 = 0$

$\text{cosec}\,x = -\frac{8}{5}$

$\sin x = -\frac{5}{8}$

$\sin^{-1}(-\frac{5}{8}) = -38.7°$

$x = 180° + 38.7°$ or $360° - 38.7°$, i.e. $x = 218.7°$ or $321.3°$ (to 1 d.p.)

b) $\sec x \quad 2\cos x = 1$

$\frac{1}{\cos x} - 2\cos x = 1$

$1 - 2\cos^2 x = \cos x$
$2\cos^2 x + \cos x - 1 = 0$
$(\cos x + 1)(2\cos x - 1) = 0$
$\cos x = -1$ or $\cos x = \frac{1}{2}$
$x = 180°$ or $x = 60°$ or $300°$

4) a) Start with $y = \sin x$, then $y = \sin 2x$ (the 2 doubles the frequency) [1 mark]
Now $y = \text{cosec}\,2x$ [1 mark]
Then shift it down 1 to give $y = \text{cosec}\,2x - 1$ [1 mark]

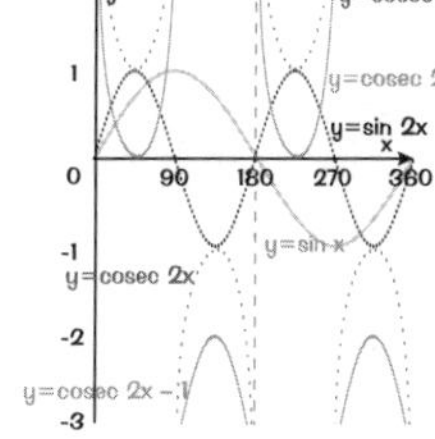

Answers

b) $\sin 2x = \text{cosec } 2x - 1$

$\sin 2x = \frac{1}{\sin 2x} - 1$ [1 mark]

$\sin^2 2x = 1 - \sin 2x$

$\sin^2 2x + \sin 2x - 1 = 0$ [1 mark]

$\sin 2x = 0.618$ or -1.618 (no solutions for this second one) [1 mark]

$2x = 38.17$ or $180 - 38.17$, or $360 + 38.17$, or $540 - 38.17$

$2x = 38.17, 141.83, 398.17$ or 501.83 [2 marks]

$x = 19.1°, 70.9°, 199.1°$ or $250.9°$ [2 marks]

Page 69

1) a) $\cot x = 3 \sin x$

$\frac{\cos x}{\sin x} = 3 \sin x$

$\cos x = 3 \sin^2 x$

$\cos x = 3(1 - \cos^2 x)$

$3\cos^2 x + \cos x - 3 = 0$

$\cos x = 0.847$ or -1.180 (no valid solutions for this second one — good)

$x = 32.1$ or $360 - 32.1$

$x = 32.1°$ or $327.9°$

b) $\text{cosec}^2 x = 2\cot x + 3$

$1 + \cot^2 x = 2 \cot x + 3$

$\cot^2 x - 2 \cot x - 2 = 0$

$\cot x = 2.73$ or -0.732

$\tan x = \frac{1}{\cot x} = 0.366$ or -1.366

$x = 20.1$ or $180 + 20.1$ or $x = -53.8$, $180 - 53.8$ or $360 - 53.8$

$x = 20.1°$ or $200.1°$ or $x = 126.2°$ or $306.2°$

2) a) $\cot^2 x = \text{cosec } x$

$\text{cosec}^2 x - 1 = \text{cosec } x$

$\text{cosec}^2 x - \text{cosec } x - 1 = 0$

$\text{cosec } x = 1.618$ or -0.618 (no valid solutions for the last one)

$\sin x = \frac{1}{\text{cosec } x} = 0.618$

Remember to put your calculator in radian mode

$x = 0.666$ rad or $\pi - 0.666$ rad

$x = 0.67$ rad or 2.48 rad to 2 d.p.

b) $\tan^2 x = 5 + \sec x$

$\sec^2 x - 1 = 5 + \sec x$

$\sec^2 x - \sec x - 6 = 0$

$(\sec x - 3)(\sec x + 2) = 0$

$\sec x = 3$ or $\sec x = -2$

$\cos x = \frac{1}{3}$ or $\cos x = -\frac{1}{2}$

$x = \pm 1.23$ rad or $x = \pm 2.09$ rad

3) a) To prove $\tan^2 x - \sec x \tan x + 1 = \frac{1}{1 + \sin x}$

Use $\tan^2 x + 1 = \sec^2 x$: LHS $= \sec^2 x - \sec x \tan x$

RHS needs only sinx, so it makes sense to turn everything into sin and cos

$\frac{1}{\cos^2 x} - \frac{1}{\cos x}\frac{\sin x}{\cos x}$ [2 marks]

$= \frac{1 - \sin x}{\cos^2 x}$ (put onto a common denominator) [1 mark]

$= \frac{1 - \sin x}{1 - \sin^2 x}$ (use $\sin^2 x + \cos^2 x = 1$ on the denominator) [1 mark]

$= \frac{1 - \sin x}{(1 + \sin x)(1 - \sin x)}$

(use difference of squares to factorise the denominator) [1 mark]

$= \frac{1}{1 + \sin x}$ = RHS [1 mark]

b) $\tan^2 x - \sec x \tan x = 1$

$\tan^2 x - \sec x \tan x + 1 = 2$

[1 mark for making use of the identity proved in part a)]

$\frac{1}{1 + \sin x} = 2$

$1 = 2 + 2\sin x$ [1 mark]

$-1 = 2\sin x$

$\sin x = -\frac{1}{2}$ [1 mark]

$x = -\frac{\pi}{6}, \pi + \frac{\pi}{6}$ or $2\pi - \frac{\pi}{6}$

$x = \frac{7\pi}{6}$ or $\frac{11\pi}{6}$ [2 marks for the solution]

Page 71

1) $\tan 15° = \tan(45° - 30°)$ (Or you could use $60° - 45°$)

$= \frac{\tan 45 - \tan 30}{1 + \tan 45 \tan 30}$ (use the identity for tan (A–B))

$= \frac{1 - \frac{1}{\sqrt{3}}}{1 + 1 \times \frac{1}{\sqrt{3}}}$

$= \frac{\sqrt{3} - 1}{\sqrt{3} + 1}$ (multiply top and bottom by $\sqrt{3}$)

$= \frac{\sqrt{3} - 1}{\sqrt{3} + 1} \times \frac{\sqrt{3} - 1}{\sqrt{3} - 1}$ (rationalise the denominator)

$= \frac{4 - 2\sqrt{3}}{2}$

$= 2 - \sqrt{3}$

2) You have a choice of three identities to use for cos 2x. Since the other function in the equation is sin x, choose the identity that involves sin x only.

$\sin x - \cos 2x = 1$

$\sin x - (1 - 2\sin^2 x) = 1$

$\sin x - 1 + 2\sin^2 x = 1$

$2\sin^2 x + \sin x - 2 = 0$

$\sin x = 0.7808$ or $\sin x = -1.28$

$x = 51.3°$ or $128.7°$

3) $4 \sin^2 \frac{x}{2} + \cos^2 x = 3$

Ugh—that $\frac{x}{2}$ looks horrid. But use the half-angle identity

$\sin^2 \frac{x}{2} = \frac{1}{2}(1 - \cos x)$ and it turns into an ordinary quadratic.

Make sure that your calculator is in radian mode.

$4 \times \frac{1}{2}(1 - \cos x) + \cos^2 x = 3$

$2(1 - \cos x) + \cos^2 x = 3$

$2 - 2\cos x + \cos^2 x = 3$

$\cos^2 x - 2\cos x - 1 = 0$

$\cos x = 2.414$ (no valid solutions – good)

or $\cos x = -0.4142$ so $x = \pm 2.00$ radians

4) a) To prove $\frac{1 - \tan^2 x}{1 + \tan^2 x} = \cos 2x$

LHS $= \frac{1 - \tan^2 x}{1 + \tan^2 x}$

$= \frac{1 - \tan^2 x}{\sec^2 x}$ [1 mark]

$= (1 - \tan^2 x)\cos^2 x$ [1 mark]

$= \cos^2 x - \cos^2 x \frac{\sin^2 x}{\cos^2 x}$ [1 mark]

$= \cos^2 x - \sin^2 x$

$= \cos 2x$ [1 mark]

= RHS

b) $\tan x \tan 2x = 1$

$\tan x \frac{2\tan x}{1 - \tan^2 x} = 1$ [1 mark]

$2 \tan^2 x = 1 - \tan^2 x$ [1 mark]

$3 \tan^2 x = 1$

$\tan^2 x = \frac{1}{3}$

$\tan x = \pm\frac{1}{\sqrt{3}}$ [1 mark]

$x = \frac{\pi}{6}, \frac{5\pi}{6}, \frac{7\pi}{6}$ or $\frac{11\pi}{6}$ [2 marks]

Answers

Page 73

1) a) $5\cos\theta + 3\sin\theta = R\cos(\theta - \alpha)$
$= R\cos\theta \cos\alpha + R\sin\theta \sin\alpha$
so $R\cos\alpha = 5$
$R\sin\alpha = 3$
Squaring and adding gives:
$R^2\cos^2\alpha + R^2\sin^2\alpha = 25 + 9$
$R^2(\cos^2\alpha + \sin^2\alpha) = 34$
$R^2 = 34$ (since $\cos^2\alpha + \sin^2\alpha = 1$)
$R = \sqrt{34}$
Dividing the equations gives:

$\tan\alpha = \frac{3}{5}$

$\alpha = 31.0°$
Putting these together gives $5\cos\theta + 3\sin\theta = \sqrt{34}\cos(\theta - 31.0°)$

b) *Max value of* $5\cos\theta + 3\sin\theta$ = *max value of* $\sqrt{34}\cos(\theta - 31.0)$
max = $\sqrt{34}$ *and occurs when* $\cos(\theta - 31.0) = 1$
when $(\theta - 31.0) = 0$
$\theta = 31.0°$
Min value of $5\cos\theta + 3\sin\theta$ = *min value of* $\sqrt{34}\cos(\theta - 31.0)$
$= -\sqrt{34}$ *and occurs when* $\cos(\theta - 31.0) = -1$
when $(\theta - 31.0) = 180$
$\theta = 180 + 31.0 = 211.0°$

2) $6\sin\theta - 5\cos\theta = R\sin(\theta - \alpha) = R\sin\theta \cos\alpha - R\cos\theta \sin\alpha$
so $R\cos\alpha = 6$ and $R\sin\alpha = 5$
Squaring and adding gives:
$R^2\cos^2\alpha + R^2\sin^2\alpha = 36 + 25$
$R^2 = 61$
$R = \sqrt{61}$

Dividing gives: $\tan\alpha = \frac{5}{6}$ $\alpha = 0.6947$

This gives $6\sin\theta - 5\cos\theta = \sqrt{61}\sin(\theta - 0.6947)$
Solving $6\sin\theta - 5\cos\theta = 4$ *is now the same as solving*
$\sqrt{61}\sin(\theta - 0.6947) = 4$

$\sin(\theta - 0.6947) = \frac{4}{\sqrt{61}}$

$\theta - 0.6947 = 0.5377$ or $\pi - 0.5377$
$\theta = 1.23$ or 3.30 *radians to 3 s.f.*

3) a) *The minimum value of* $\frac{1}{f(x)}$ *will occur when f(x) has its maximum value, so start by finding that.*
$f(x) = 9\cos x - 3\sin x + 5$ (*you can ignore the + 5 for the moment.*)
$9\cos x - 3\sin x = R\cos(x + \alpha)$
[1 mark for realising that you need $R\cos(x + \alpha)$*]*
$= R\cos x \cos\alpha - R\sin x \sin\alpha$
so $R\cos\alpha = 9$ and $R\sin\alpha = 3$
Squaring and adding gives:
$R^2 = 81 + 9 = 90$
$R = \sqrt{90}$

Dividing gives $\tan\alpha = \frac{3}{9} = \frac{1}{3}$ $\alpha = 18.43°$ *[3 marks for R and* α*]*

$9\cos x - 3\sin x = \sqrt{90}\cos(x + 18.43)$
The max value of this is $\sqrt{90}$
so the max value of f(x) is $\sqrt{90} + 5$. *[1 mark for max value]*

so the min value of $\frac{1}{f(x)}$ *is* $\frac{1}{(\sqrt{90}+5)} = 0.07$ *to 2 d.p.*

[1 mark for turning it into the min value that you need]
and occurs when $\cos(x + 18.43) = 1$
when $x + 18.43 = 0$ *or 360 to give the smallest positive value you need*
$x = 360 - 18.43 = 341.57°$ *to 2 d.p. [1 mark]*

b) $f(x) = 3$
$9\cos x - 3\sin x + 5 = 3$
$9\cos x - 3\sin x = -2$
$\sqrt{90}\cos(x + 18.43) = -2$ *[1 mark]*

$\cos(x + 18.43) = \frac{-2}{\sqrt{90}}$

$x + 18.43 = 102.17$ *or* $360 - 102.17$ *[1 mark]*
$x = 84°$ *or* $239°$ *to nearest degree [2 marks]*

Page 75

1) *To prove* $2\cos 3x \sin x = \sin 2x(1 - 4\sin^2 x)$
LHS = $\sin(3x + x) - \sin(3x - x)$
(*using* $2\cos A \sin B = \sin(A + B) - \sin(A - B)$)
$= \sin 4x - \sin 2x$
$= 2\sin 2x \cos 2x - \sin 2x$ (*using* $\sin 2x = 2\sin x \cos x$...
...except that this time you want to turn $\sin 4x$ *into* $2\sin 2x \cos 2x$)
$= \sin 2x(2\cos 2x - 1)$
$= \sin 2x(2(1 - 2\sin^2 x) - 1)$ (*using* $\cos 2x = 1 - 2\sin^2 x$)
$= \sin 2x(1 - 4\sin^2 x)$

2) $\sin 3x - \sin x = 0$

Use $\sin X - \sin Y = 2\cos\frac{X+Y}{2}\sin\frac{X-Y}{2}$

$2\cos 2x \sin x = 0$
$\cos 2x = 0$ *or* $\sin x = 0$

$2x = \frac{\pi}{2}, \frac{3\pi}{2}, \frac{5\pi}{2}$ *or* $\frac{7\pi}{2}$ *or* $x = 0, \pi$ *or* 2π

So the solutions are: $x = 0, \frac{\pi}{4}, \frac{3\pi}{4}, \pi, \frac{5\pi}{4}, \frac{7\pi}{4}, 2\pi$

3) $$\tan\left(\frac{\pi}{4}+\theta\right) = \frac{\sin\left(\frac{\pi}{4}+\theta\right)}{\cos\left(\frac{\pi}{4}+\theta\right)} = \frac{\sin\frac{\pi}{4}\cos\theta + \cos\frac{\pi}{4}\sin\theta}{\cos\frac{\pi}{4}\cos\theta - \sin\frac{\pi}{4}\sin\theta} = \frac{\frac{1}{\sqrt{2}}\cos\theta + \frac{1}{\sqrt{2}}\sin\theta}{\frac{1}{\sqrt{2}}\cos\theta - \frac{1}{\sqrt{2}}\sin\theta}$$

$$\approx \frac{\frac{1}{\sqrt{2}}\left(1 - \frac{\theta^2}{2} + \theta\right)}{\frac{1}{\sqrt{2}}\left(1 - \frac{\theta^2}{2} - \theta\right)} \approx \frac{1+\theta}{1-\theta} \text{ if } \theta^2 \text{ can be ignored.}$$

4) a) *To prove* $\frac{\cos y - \cos x}{\sin x - \sin y} = \tan\frac{x+y}{2}$

LHS $= \frac{\cos y - \cos x}{\sin x - \sin y}$

$= \frac{-2\sin\frac{x+y}{2}\sin\frac{y-x}{2}}{2\cos\frac{x+y}{2}\sin\frac{x-y}{2}}$ (*using the sum and product identities*) *[2 marks]*

$= \frac{2\sin\frac{x+y}{2}\sin\frac{x-y}{2}}{2\cos\frac{x+y}{2}\sin\frac{x-y}{2}}$ (*using* $-\sin x = \sin(-x)$) *[1 mark]*

$= \frac{\sin\frac{x+y}{2}}{\cos\frac{x+y}{2}}$ *[1 mark for cancelling]*

$= \tan\frac{x+y}{2}$ = RHS *[1 mark]*

b) $\frac{\cos x - \cos 3x}{\sin 3x - \sin x} + \tan x = 0$
Use the result proved in part a):

$\tan\frac{x+3x}{2} + \tan x = 0$

[1 mark for trying to use it, 1 mark for getting it right]
$\tan 2x + \tan x = 0$

$\frac{2\tan x}{1 - \tan^2 x} + \tan x = 0$ *[1 mark for the identity of tan 2x]*
$2\tan x + \tan x(1 - \tan^2 x) = 0$ *[1 mark for rearranging the equation]*
$3\tan x - \tan^3 x = 0$
$\tan x(3 - \tan^2 x) = 0$ *[1 mark for factorising]*
$\tan x = 0$ *or* $\tan x = \pm\sqrt{3}$

$x = 0, \pi$ *or* 2π *or* $x = \frac{\pi}{3}, \frac{2\pi}{3}, \frac{4\pi}{3}$ *or* $\frac{5\pi}{3}$ *[3 marks]*

Index